Innovation and the Productivity Crisis

THE BROOKINGS INSTITUTION is an independent organization devoted to nonpartisan research, education, and publication in economics, government, foreign policy, and the social sciences generally. Its principal purposes are to aid in the development of sound public policies and to promote public understanding of issues of national importance.

The Institution was founded on December 8, 1927, to merge the activities of the Institute for Government Research, founded in 1916, the Institute of Economics, founded in 1922, and the Robert Brookings Graduate School of Economics and Government, founded in 1924.

The Board of Trustees is responsible for the general administration of the Institution, while the immediate direction of the policies, program, and staff is vested in the President, assisted by an advisory committee of the officers and staff. The by-laws of the Institution state: "It is the function of the Trustees to make possible the conduct of scientific research, and publication, under the most favorable conditions, and to safeguard the independence of the research staff in the pursuit of their studies and in the publication of the results of such studies. It is not a part of their function to determine, control, or influence the conduct of particular investigations or the conclusions reached."

The President bears final responsibility for the decision to publish a manuscript as a Brookings book. In reaching his judgment on the competence, accuracy, and objectivity of each study, the President is advised by the director of the appropriate research program and weighs the views of a panel of expert outside readers who report to him in confidence on the quality of the work. Publication of a work signifies that it is deemed a competent treatment worthy of public consideration but does not imply endorsement of conclusions or recommendations.

The Institution maintains its position of neutrality on issues of public policy in order to safeguard the intellectual freedom of the staff. Hence interpretations or conclusions in Brookings publications should be understood to be solely those of the authors and should not be attributed to the Institution, to its trustees, officers, or other staff members, or to the organizations that support its research.

Innovation and the Productivity Crisis

Martin Neil Baily
Alok K. Chakrabarti

The Brookings Institution
Washington, D.C.

Copyright © 1988 by

THE BROOKINGS INSTITUTION

1775 Massachusetts Avenue, N.W., Washington, D.C. 20036

Library of Congress Cataloging-in-Publication Data:

Baily, Martin Neil.
    Innovation and the productivity crisis / Martin Neil Baily and
Alok K. Chakrabarti.
        p. cm.
    Includes bibliographical references and index.
    ISBN 0-8157-0760-6 ISBN 0-8157-0759-2 (pbk.)
    1. Industrial productivity. 2. Technological innovations—
Economic aspects. I. Chakrabarti, Alok K. II. Title.
HC79.I52B35 1988   338′ .06—dc 19   88-1697
                                                          CIP

9 8 7 6 5 4 3 2 1

Foreword

THE SLOW GROWTH of productivity in the U.S. economy over the past fifteen years has presented a puzzle for analysts and a challenge for economic policy. American living standards have improved somewhat during this period, but largely because of an increase in employment and the resort to foreign borrowing. If living standards are to continue to rise, however, productivity growth must contribute more in the future.

In this book, Martin Neil Baily and Alok K. Chakrabarti present a profile of the slowdown in productivity growth and review the many contributory elements. They find the standard explanations leave a large gap that strongly suggests productivity growth may have been pulled down by the slow pace of innovation. Evidence from several case studies reveals that, although opportunities for technological advance were available, U.S. industry either missed them or was unable to use new technologies in ways that raised productivity. To encourage innovation, the authors propose enactment of an enhanced tax credit for research and development. They also advocate encouraging coalitions of private companies to perform middle-ground research—research oriented to commercial applications but too general to justify investment by individual companies. They recommend both easing government obstacles to joint ventures and providing some public cofunding if the projects promise social returns.

This book continues the tradition of productivity studies at Brookings pioneered by Edward Denison's analyses of the determinants of productivity growth both here and overseas. Financial support was provided by the Chase Manhattan Foundation and by the Division of Science Resource Studies of the National Science Foundation through grants to Drexel University.

Martin Neil Baily is a senior fellow in the Economic Studies program at the Brookings Institution and Alok K. Chakrabarti is associate vice president for research and William A. Mackie Professor of Commerce and Engineering at Drexel University. They would like to thank Barry P. Bosworth, Robert W. Crandall, Michael F. Mohr, Richard R. Nelson, and

v

F. M. Scherer for many helpful comments. Nathaniel Levy, Gregory I. Hume, Robert Krebs, and Hiranthi D. R. de Silva provided research assistance. Jeanette Morrison edited the manuscript, Almaz S. Zelleke verified its factual content, and Diana Regenthal prepared the index. Sara C. Hufham provided secretarial support. Their assistance is gratefully acknowledged.

The views expressed here are those of the authors and should not be ascribed to those acknowledged above or to the trustees, officers, or other staff members of the Brookings Institution.

BRUCE K. MACLAURY
President

March 1988
Washington, D.C.

Contents

Tables

Figures

CHAPTER ONE

The Productivity Crisis

PRODUCTIVITY GROWTH in the U.S. economy has collapsed in recent years, and the size of the collapse has the dimensions of a crisis. In the private business sector, output per hour of labor (one standard measure of productivity) grew 3.3 percent a year between 1948 and 1965 but slumped to 1.4 percent a year between 1965 and 1985. Had growth continued at its pre-1965 rate, U.S. output in 1985 would have been 45 percent higher than it actually was, with no additional labor used in production.

Because the growth rate from 1948 to 1965 was unusually high by historical standards, perhaps it is unrealistic to expect that the postwar boom could have continued indefinitely. But using more traditional rates still shows the effect of the dramatic slowdown. We have calculated that, had the growth in output per hour after 1965 simply equaled the average rate from 1870 to 1965, output would still have been more than 20 percent higher in 1985 than it actually was.[1]

The Dimensions of the Problem

To put the decline in perspective, this amount of additional output is larger than that needed to solve many of today's economic problems, notably the budget deficit. Productivity is the mechanism by which economic resources expand. Unless there is an increase in growth, Americans will suffer stagnant or falling living standards. Output per hour of labor in the business sector is strongly linked to real wages, the key determinant of

1. Our calculations are based on data from Angus Maddison, "The Productivity Slowdown in Historical and Comparative Perspective," University of Groningen, the Netherlands, 1985. This paper has been published in slightly different form as "Growth and Slowdown in Advanced Capitalist Economies: Techniques of Quantitative Assessment," in *Journal of Economic Literature*, vol. 25 (June 1987), pp. 649–98.

living standards for most Americans. After 1965 compensation per hour grew at only half the rate it had achieved before then. From 1973 to 1985 real compensation per hour grew at only 0.8 percent a year.[2]

In addition to output per hour worked, or labor productivity, there are two other common yardsticks used to track productivity performance: output per unit of capital used and multifactor productivity (MFP). Table 1-1 shows the growth in all three productivity measures for the private business sector, the nonfarm business sector, and the manufacturing sector of the U.S. economy for several periods since 1948.

The middle column reveals that the owners of capital, like the workers discussed above, did poorly after 1965. Although output per unit of capital does not have a pronounced historical trend, the ratio rose 13 percent for the nonfarm business sector between 1948 and 1965 only to then fall 22 percent through 1986.[3] As figure 1-1 confirms, the rate of profit declined substantially after 1965.

While the single-factor measures are important for certain types of comparisons, multifactor productivity growth is preferable to either as an indicator of overall economic performance because it measures the efficiency with which both capital and labor together are used in production. Briefly, it is the ratio of output to a weighted average of the capital and labor inputs, where the weights are the relative importance of the two factors in the cost of production.[4]

The collapse of multifactor productivity growth in the aggregate is startling. For nearly a decade and a half (1973–86) there was little increase in multifactor productivity in the private nonfarm business sector of the U.S. economy. Moreover, this crisis shows no clear sign of abating: the performance between 1979 and 1986 in the aggregate was almost as bad as that between 1973 and 1979.

2. *Economic Report of the President, January 1987*, pp. 248, 294. The compensation figures are for the business sector and include fringe benefits. The adjustment for inflation was made using the deflator for personal consumption expenditure. By some other measures wages were even worse. Average hourly earnings in the private nonfarm sector were actually down 7 percent in 1985 compared with 1973. These data exclude fringe benefits and are adjusted for inflation using the consumer price index. Ibid., p. 292.

3. U.S. Bureau of Labor Statistics, "Multifactor Productivity Measures, 1986," *News*, USDL 87-436, October 13, 1987, table 2.

4. Detailed information on the calculation of multifactor productivity is given in Bureau of Labor Statistics, *Trends in Multifactor Productivity, 1948–81*, Bulletin 2178 (Washington, D.C.: U.S. Department of Labor, September 1983). In the analysis of productivity by industry, multifactor productivity is often computed using material and energy inputs as separate factors of production.

Table 1-1. *Three Measures of Productivity Growth in the U.S. Business, Nonfarm Business, and Manufacturing Sectors, 1948–86*[a]
Average annual percentage change

Period	Productivity Measure		
	Output per hour of labor	Output per unit of capital	Multifactor productivity growth
Business sector			
1948–65	3.25	0.81	2.39
1965–73	2.14	−0.69	1.12
1973–79	0.61	−0.88	0.10
1979–86	1.40	−1.05	0.52
Nonfarm business sector[b]			
1948–65	2.70	0.77	2.02
1965–73	1.83	−0.79	0.91
1973–79	0.48	−1.14	−0.08
1979–86	1.16	−1.28	0.33
Manufacturing sector			
1948–65	2.92	0.78	2.26
1965–73	2.48	−0.91	1.46
1973–79	1.37	−1.85	0.52
1979–86	3.42	−0.06	2.53

Source: U.S. Bureau of Labor Statistics, "Multifactor Productivity Measures, 1986," *News,* USDL 87-436, October 13, 1987, tables 1, 2, 3.
a. Calculated using natural logs.
b. Major components are manufacturing, mining, construction, transportation, public utilities, wholesale and retail trade, finance and insurance, real estate, and services. Excludes government operations, agriculture, and nonprofit organizations.

The Slowdown in Other Countries

Output per hour (average labor productivity) is the most available productivity measure for international comparisons. Table 1-2 shows the growth of gross domestic product (GDP) for each hour worked in five countries over the period 1950–84. (Because it includes government operations, GDP is a more comprehensive concept of output than those used in table 1-1.) The table also shows how much output per hour in manufacturing grew in those countries as calculated by the U.S. Bureau of Labor Statistics.

Clearly the United States is not alone in experiencing a decline in productivity growth. The slowdown occurred in each of the industrialized countries; in fact most of them have slowed more than has the United States. This book is about the U.S. economy, but it is still important to keep the international situation in mind. Explanations of the slowdown that

Figure 1-1. *Rate of Return on Tangible Assets: Actual and Cyclically Adjusted Values, 1950–80*[a]

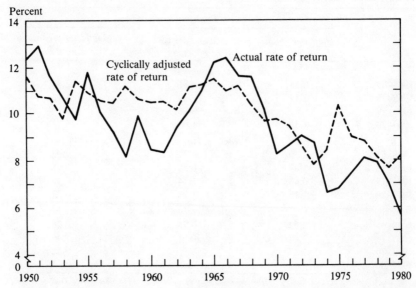

Percent

Source: Barry P. Bosworth, "Capital Formation and Economic Policy," *Brookings Papers on Economic Activity, 2:1982,* fig. 3.
a. Tangible assets include equipment, structures, land, and inventories.

cite factors particular to the United States must be suspect. For example, the 55-mph speed limit might have pulled down productivity here, but could hardly have done so in Germany.

Productivity Growth by Industry

Table 1-3 shows productivity growth in the major manufacturing and nonmanufacturing industries of the U.S. economy.[5] These data establish a striking fact: the slowdown is pervasive across the U.S. economy. Between 1973 and 1979 fourteen out of twenty manufacturing industries and six out of nine nonmanufacturing industries grew more slowly than their average rate for the whole period 1948–85 (with a difference exceeding half a percentage point). Matters did not improve after 1979; the cor-

5. These data were prepared by the American Productivity Center using slightly different methods than the Bureau of Labor Statistics (BLS) figures in table 1-1, but the differences are not great enough to change the overall picture.

Table 1-2. *Labor Productivity Growth in Five Countries, Aggregate and Manufacturing, Selected Periods, 1950–86*[a]
Average annual percentage change

Period	France	Germany	Japan	Britain	United States
	Growth of GDP per hour worked				
1950–73	5.01	5.83	7.41	3.15	2.44
1973–79	3.83	3.91	3.40	2.18	0.80
1979–84	3.24	1.88	3.06	2.95	1.09
	Growth of manufacturing output per hour				
1950–73	5.63	6.31	9.48	3.25	2.62
1973–79	4.90	4.22	5.39	1.15	1.37
1979–86	3.50	2.78	5.46	4.28	3.10

Sources: For GDP per hour: Angus Maddison, "The Productivity Slowdown in Historical and Comparative Perspective," University of Groningen, the Netherlands, 1985, table A-5. For manufacturing: Bureau of Labor Statistics, "International Comparisons of Manufacturing Productivity and Labor Cost Trends, 1986," *News,* USDL 87-237, June 15, 1987.
a. Calculated using natural logs.

responding figures for 1979–85 are fourteen out of twenty manufacturing industries and eight out of nine nonmanufacturing industries.

Looking at the productivity slowdowns by industry fails to reveal any obvious patterns. Some old-line manufacturing industries, such as steel, have done poorly, while others, such as textiles and apparel, have done well. The service sector has never had much growth, but it has not slowed down. Finance and insurance, a dynamic industry, has done terribly. Especially conspicuous for their negative growth are mining and construction; construction in particular dragged down aggregate productivity after 1965. Based on the data in table 1-3, the *level* of multifactor productivity in construction in 1985 was well below the level achieved in 1948.

There is one bright spot in the productivity series: the performance of the manufacturing sector. Productivity growth in manufacturing in 1979–85 resumed its normal postwar rate. This exception to the prevailing trend is important but needs to be understood in relation to three facts. First, the manufacturing recovery is heavily tied to the performance of a single industry, computers (part of nonelectrical machinery, standard industrial classification 35). Removing SIC 35 data from the manufacturing sector lowers the rate of multifactor productivity growth for 1979–85 a full percentage point.[6] Second, the manufacturing sector was exposed to recession in 1980–82 and then to extreme foreign competition

6. Calculated by the authors using data from the American Productivity Center and the U.S. Department of Commerce, Bureau of Economic Analysis. Edward Denison uses price data to remove just computers from the manufacturing series and finds the same 1 percentage point drop that we found by removing nonelectrical machinery. Personal communication from Edward F. Denison, Senior Fellow Emeritus, Brookings Institution.

Table 1-3. *Multifactor Productivity Growth in Major U.S. Industries, 1948–85*

Average annual percentage change

Industry	1948–85	1948–65	1965–73	1973–79	1979–85
Major aggregates					
Farming	3.4	3.4	2.8	1.8	5.6
Manufacturing	2.0	2.5	1.8	0.6	2.1
Nonfarm nonmanufacturing	1.1	2.1	0.9	−0.2	−0.4
Manufacturing except nonelectrical machinery[a]	1.9	2.6	1.9	0.6	1.1
Manufacturing industries					
Food	2.6	2.9	3.6	0.1	2.9
Tobacco	0.2	2.7	2.0	−0.2	−8.0
Textiles	3.9	4.6	2.0	5.7	2.6
Apparel	2.1	1.9	2.7	2.8	1.5
Lumber	2.4	3.8	1.0	1.5	1.3
Furniture	1.7	2.0	1.6	2.1	0.5
Paper	2.0	2.0	3.7	0.1	1.8
Printing and publishing	1.0	2.1	0.8	−0.2	−0.9
Chemicals	3.2	4.1	4.0	1.1	2.0
Petroleum	0.6	2.9	0.8	−1.4	−4.0
Rubber	1.9	2.2	1.7	−0.8	4.0
Leather	0.8	1.1	1.8	0.2	−0.6
Stone, clay, and glass	1.4	2.2	1.0	0.5	0.9
Primary metals	0.0	1.0	0.7	−2.8	−0.9
Fabricated metals	1.3	1.8	1.3	0.0	1.1
Nonelectrical machinery	2.5	1.4	1.5	0.4	9.1
Electrical machinery	3.4	4.2	2.7	3.5	1.9
Transportation equipment	2.0	3.5	1.2	0.1	0.6
Instruments	2.5	3.6	2.4	2.7	−0.5
Miscellaneous manufactures	2.0	2.6	2.9	0.1	1.2
Nonmanufacturing industries					
Mining	0.6	3.1	2.0	−6.0	−1.5
Construction	−0.2	2.9	−3.9	−2.2	−2.0
Transportation	1.6	2.1	2.7	1.5	−1.3
Railroad	2.7	4.1	1.8	1.3	1.4
Nonrail	0.7	0.5	2.7	1.3	−2.0
Communications	3.9	5.6	3.4	2.4	1.3
Public utilities	3.5	5.8	3.0	−0.6	1.7
Trade	1.9	2.6	2.4	0.4	0.8
Finance and insurance	0.3	1.3	0.6	−0.7	−2.0
Real estate	0.6	1.8	0.3	1.4	−3.2
Services	0.7	0.4	1.4	0.4	0.8

Source: American Productivity Center, *Multiple Input Productivity Indexes*, vol. 6 (December 1986), table 3a, calculated using data from the U.S. Bureau of Labor Statistics and the U.S. Department of Commerce, Bureau of Economic Analysis.

a. Authors' calculations.

in 1981–85 as the dollar rose. These conditions forced many companies to reorganize and to close inefficient plants. The resulting productivity gains are real but do not ensure that productivity will continue to grow.

Third, manufacturing may have done worse, and nonmanufacturing better, than the table shows. Contrary to a widespread impression, analysts measuring productivity do not start with industry data and work up to the aggregate. Instead, they measure changes in aggregate output, then allocate the growth by industry. This procedure has the potential for serious error. For example, before the gross national product accounts were revised in 1982, some observers thought productivity growth was being understated in manufacturing and overstated in nonmanufacturing because no adequate price index existed for computers. The estimated real output of computers was too low, but the real capital stocks of industries buying computers were also too low, so their estimated multifactor productivity growth rates were too high.

Although this error has been corrected, there may have been offsetting errors in the opposite direction that have not been corrected. One piece of evidence is that analysts calculating multifactor productivity growth in the nonmanufacturing sector have come up with negative figures since 1973. That result is pretty implausible. Edward Denison argues persuasively that one can have much more confidence in estimates of aggregate output and aggregate inputs than in allocations of these aggregates by industry, in part because final goods prices are surveyed more accurately than are prices of intermediate goods.[7]

Computers and Productivity

Valuing computers and estimating the performance of computer-using industries raise issues whose importance goes well beyond measurement problems. During the 1970s and 1980s computers and electronics have made up a major locus of rapid growth and innovation within the U.S. economy, and the electronics revolution has transformed the office and the factory in many ways. But paradoxically, so far there has been little sign of the payoff to productivity among the industries buying the equipment produced.[8] That is a major puzzle we explore in this book.

7. Edward F. Denison, "Progress in Productivity Research," in Reino T. Hjerppe, Aarno Laihonen, and Pertti Parmanne, eds., *From Tabulation to Information Society* (Helsinki, Finland: Central Statistical Office of Finland, 1986), p. 34.
8. The communications industry is a major exception to this statement.

Productivity and Competitiveness

The huge deficits in the U.S. trade account and current account have made competitiveness an important policy issue. And some argue that slow productivity growth in the United States is a key cause of the trade deficit. In common with most economists, we think that the overvaluation of the dollar from 1981 to 1985 was the major reason for the trade deficit, and that the federal budget deficit was the main reason for the overvaluation. Improvements in U.S. productivity growth are desirable, but they are neither necessary nor sufficient for obtaining external economic balance. The dollar has now fallen and, provided it stays down, the foreign account should improve.

Some concerns about U.S. competitiveness go beyond the short-run trade deficit, however. Although all the industrial countries experienced slowdowns in growth after 1973, many of them are still achieving productivity increases much more rapid than those in the United States. For example, between 1973 and 1986 output per hour in manufacturing grew 3.1 percent a year faster in Japan than it did in the United States. The corresponding figures were 1.8 percent faster in France and 1.1 percent faster in Germany.[9] The question is being asked, Will the United States be overtaken economically?

In addition, it has been found that the value of the dollar required to maintain external balance for the United States has declined over time. This means that the U.S. terms of trade are deteriorating, so that a given volume of exports can be exchanged for a declining volume of imports. It is vis-à-vis Japan that the dollar decline shows up most strongly. Between 1967 and 1987 there was about as much inflation in Japan as in the United States. And yet during that time the yen rose against the dollar by a factor of 2.6.

In earlier periods the United States was able to maintain trade balance when the dollar was strong because it produced many unique goods and sold them for premium prices. But Americans have lost much of the technological lead that allowed this favorable state of affairs, and U.S. exporters now face a more competitive environment.

Table 1-4 shows how the levels of productivity in other countries have compared with the level in the United States since 1950. Output per hour

9. Bureau of Labor Statistics, "International Comparisons of Manufacturing Productivity and Labor Cost Trends, 1986," *News*, USDL 87-237, June 15, 1987. Calculated using natural logs.

Table 1-4. *Average Labor Productivity in Four Countries Relative to the United States, Aggregate and Manufacturing, Selected Years, 1950–86*

Year	France	Germany	Japan	Britain	United States
	Aggregate economy (GDP per hour worked)				
1950	39.6	31.5	13.6	55.9	100
1960	47.5	47.9	18.4	54.8	100
1973	71.4	68.7	42.6	65.8	100
1980	85.8	83.7	51.5	71.7	100
1984	95.3	86.1	54.9	78.4	100
	Manufacturing output per hour				
1950	32.9	28.6	11.4	36.3	100
1960	43.8	48.1	23.2	36.6	100
1973	65.8	66.8	55.3	41.9	100
1980	82.6	79.6	75.2	41.1	100
1984	85.8	78.7	81.3	45.5	100
1985	85.0	78.7	83.6	45.2	100
1986	83.6	77.5	83.0	44.9	100

Sources: For GDP per hour: Maddison, "Productivity Slowdown," table A-5. Maddison relies heavily on data from the OECD. For manufacturing: productivity relatives for 1983 were taken from Data Resources, Inc., *Review of the U.S. Economy* (April 1986), p. 23, and then other years were computed using rate of growth from BLS, "International Comparisons of Manufacturing Productivity."

of labor is given for four major countries relative to the United States for both the aggregate economy (GDP per hour) and manufacturing. The data show that the United States remained the most productive economy in 1984. If one looks at GDP per hour, the Japanese economy in 1984 was still at a rather low level of productivity, only 55 percent of that in the United States. On the basis of recent trends, it will be well into the next century before Japan catches up. On the other hand, productivity in the French economy is high, roughly comparable to that in the United States, and German productivity is pretty close also. Again, recent trends demonstrate that productivity in France is overtaking that in the United States already, although Germany is barely closing the remaining productivity differential.[10]

The superior performance of Japan in foreign trade derives from its highly efficient manufacturing sector, one that has narrowed the gap with

10. The strength of the French economy looks greatest in productivity terms and looks less impressive in terms of output. The French worked 54 percent fewer hours per capita in 1984 than did the Japanese as a result of a shorter work-week, more vacations, a lower level of labor-force participation, and higher unemployment. The United States is an intermediate case, working 22 percent more than the French and 33 percent less than the Japanese in 1984. Authors' calculations based on Maddison, "Productivity Slowdown," tables A-2, A-3.

the United States rapidly indeed. If recent trends continue, productivity in Japanese manufacturing will overtake that in the United States by 1994. In some manufacturing industries it has already overtaken the United States.

If U.S. competitiveness is defined to include the performance of U.S. productivity relative to our competitors, then there is a basis for concern about a long-run decline in competitiveness. Some catching up of other countries after World War II was to be expected and indeed was encouraged by U.S. policies. At this time, however, an improvement in U.S. productivity growth is necessary if the United States is to remain one of the world leaders in productivity.

Conclusions and Preview of the Book

As the price of fighting inflation, the United States went through serious recessions in 1975–77 and 1980–86. On the assumptions that the natural rate of unemployment is 6 percent and that each point of excess unemployment involves a 2.5 percent loss of output (Okun's law), it follows that the mid-1970s recession cost 12 percent of output, averaging 4 percent a year. The 1980s recession cost 32 percent of output, averaging 4.6 percent a year.

The loss of output associated with the collapse of productivity growth since 1965 is much larger than the losses from these two serious recessions. Even with conservative assumptions about long-run trends, the deficiency in growth is costing us more in one year than was lost in three years during the 1975–77 recession.

A few industries seem to have escaped the slowdown, but these exceptions do not change its pervasiveness. Two-thirds or more of the major industries in the economy have had slow productivity growth. Many have even shown declines in multifactor productivity over several years.

In the 1980s rapid innovation and productivity growth in computers and electronics have been enough to generate a growth recovery in manufacturing. Without in any way dismissing this major exception to the general rule, we note that computer-using industries have yet to transform this locus of innovation into overall productivity growth for the general benefit.

Other major industrial economies have achieved higher rates of productivity growth than the United States over many years. There is concern that we are being overtaken and are losing our place as the most productive economy, and there is some basis for this concern.

This book grew out of our desire to find out what has caused the decline in productivity growth and what can be done to increase productivity. There are many villains behind a deteriorating American performance, such as inadequate or misplaced capital investment and government regulation, and we review their individual contributions. However, we find these explanations of the slowdown leave a large puzzle that strongly suggests productivity growth was pulled down by slow innovation. That view leads us to a basic question: have technological opportunities themselves declined, or have available opportunities been missed because U.S. industry failed to incorporate new technology into production effectively? Have we run out of new ideas, or have we missed taking advantage of them, and why?

Answering that question at the macroeconomic level of analysis may be impossible. Thus we undertake studies in detail of four industries: chemicals, textiles, machine tools, and electric power. Three of the four are chosen from the long list that experienced substantial productivity slowdowns in the 1970s; textiles is the exception. We also wanted a good mix of characteristics: high-tech and low-tech, high and low R&D spending, high and average capital intensity.

To add to what has been learned in the past from econometric models, we develop detailed measures of what "actually happened" to the pace and character of technical advance in those industries. Specifically, we collected files of introduced innovations by surveying the industries' periodicals, and we conducted interviews with industry specialists and managers.

Evidence from our case studies lies at the center of the book. Though eclectic, our results clearly show how the evolution and application of technological advances in these industries have been a vital part of overall performance and productivity.

Our final case, the computer and electronics industry, requires a different approach. It is self-evident that for two decades computers and electronics have been highly innovative and dynamic. What is not as obvious is how far this technology has benefited the *using* industries—what we have called the paradox of the electronics revolution. Here we use several analytical tools to examine various hypotheses, both optimistic and pessimistic, about white-collar productivity.

Looking at the evidence we conclude that, although the scientific frontier has continued to advance, some technological opportunities were indeed missed. Why they were missed, how those failures affected productivity growth, and what can be done by the public and private sectors to stimulate growth to meet the productivity challenge are questions addressed in the final chapter.

CHAPTER TWO

Explanations of the Slowdown

ECONOMISTS and others have offered numerous views on why productivity growth has slowed, but the serious explanations can be grouped into eight possibilities.[1] We first present the suspects in a lineup with brief explanations, and then go through the evidence against them in more detail.

LABOR. The growth of educational attainment may have slowed, the level of skill and experience in the labor force may have deteriorated, or workers may not work as hard as they used to.

CAPITAL. Capital investment may have been inadequate to sustain the level of productivity growth, or may not have been very productive.

ENERGY AND MATERIALS. The price of energy and other raw materials increased around the same time that productivity growth slowed. In an attempt to economize on these inputs, companies may have substituted materials for capital and labor and reduced measured multifactor productivity growth.

OUTPUT MEASUREMENT. The products and services produced by the economy are diverse and vary over time. Part of the slowdown may be a statistical illusion created by measurement problems.

THE COMPOSITION OF OUTPUT. The level of productivity differs greatly in the different industries of the economy. If production shifts toward industries with either low levels of productivity or low rates of growth, such as service activities, the shift may pull down the average growth in productivity.

MANAGEMENT FAILURES. The number of master's of business administration (MBA) holders running companies has proliferated, even as productivity growth has declined. Critics say that U.S. managers have emphasized financial manipulation and short-term paper profits at the expense of sound investments and technology development. We have managed our way to decline, it is claimed.

1. This chapter draws on Martin Neil Baily, "What Has Happened to Productivity Growth?" *Science*, vol. 234 (October 24, 1986), pp. 443–51.

GOVERNMENT REGULATORY AND DEMAND POLICIES. By imposing an increased burden of regulatory requirements and by allowing high and fluctuating rates of inflation, and low and fluctuating rates of growth of product demand, the government, it is said, has disrupted economic efficiency.

TECHNOLOGY. Because innovation is a key source of economic growth, a decline in the pace of innovation may have slowed productivity growth. That possibility is a main theme of this book, and one we will explore in this and later chapters.

Labor Quality and Effort

The simple productivity calculations given in chapter 1 are made by using worker-hours as the measure of labor input. This procedure gives the same weight to an hour of unskilled labor as to an hour by an engineer. Clearly that is not right. People differ greatly in their abilities and training. The principal way to track changes over time in the skill level of the work force is by estimating how educated it is. One economist, Michael R. Darby, argues that much of the productivity slowdown can be attributed to the allegedly slow rate at which educational attainment increased in the 1970s.[2]

Darby uses as his index of education the median years of schooling of the adult population. That number increased rapidly through the 1950s and 1960s, but stopped growing in the 1970s. In terms of correlation, therefore, it serves well as a culprit in the slowdown. Unfortunately, it is not a useful measure of educational attainment for analyzing productivity. At the end of World War II the median years of schooling was about nine: that is, half the population had had nine years of schooling or more and half had had nine years or less. In the postwar period the median rose rapidly until it hit twelve years in the late 1960s. It then stopped growing, because a large fraction of the population completes only high school. This does not mean that educational attainment actually stopped rising. In 1980, 16 percent of the population had four or more years of college, and 35 percent of the population had completed high school, whereas the corresponding figures for 1970 are 11 percent and 31 percent.[3] In other words, the time-series

2. Michael R. Darby, "The U.S. Productivity Slowdown: A Case of Statistical Myopia," *American Economic Review*, vol. 74 (June 1984), pp. 301–22.

3. U.S. Bureau of the Census, *Statistical Abstract of the United States, 1987* (Washington, D.C.: Department of Commerce, 1986), p. 121.

behavior of the median years of schooling does not correspond to the underlying reality. It is a quirk of the particular statistical measure.

The most careful work on how education affects economic growth has been done by Edward F. Denison.[4] First, he uses estimates of how much an additional year of education adds to each worker's income. These estimates put a productivity value on each year of schooling. He also compiles detailed information on the distribution of educational attainment in the work force. Then the quantity and value information are combined to form an estimate of the educational human capital of the work force.

Denison concludes that the level of educational attainment of the U.S. work force has been rising strongly since 1948 with no decline in its growth rate in the 1970s. In fact, Denison estimates that increases in educational capital added 0.40 percentage point a year to the growth of output from 1948 to 1973 and 0.47 percentage point a year from 1973 to 1982.[5] His figures, in other words, show a slight acceleration in the contribution of education after 1973.

Quality of Education

Denison found that the U.S. work force is becoming more and more educated, but in concluding that more education led to productivity growth, he assumed that the quality of education had remained the same. But what if educational quality has diminished? One possible sign of a quality loss is that Scholastic Aptitude Test (SAT) scores followed a long downtrend after 1963. The cohort of students born in 1962 and tested in 1980 had an average score on the verbal test that was 12 percent below the score of the cohort born in 1945 and tested in 1963. The corresponding decline in the mathematical test score was 7 percent.

Can this decline in SAT scores explain the slowdown? Two obstacles stand in the way of blaming the productivity slowdown on the younger generation and the disappearance of old-fashioned educational virtues. The first is that about half of the total decline in SAT scores, and three-quarters of the decline from 1963 to 1970, resulted from a broadening of the base of people taking the test.[6] As more and more students finished high school, the bottom tail of the ability distribution lengthened.

4. Edward F. Denison, *Trends in American Economic Growth, 1929–1982* (Brookings, 1985).

5. Ibid., p. 113.

6. Landon Y. Jones, "The Mystery of the Declining SAT Scores," *Princeton Alumni Weekly,* October 20, 1980, p. 24.

The second obstacle is that young people do not make up a large enough fraction of the work force to cause the fairly abrupt productivity collapse that actually took place. The SAT scores, after adjusting for the base broadening, indicate that at most a 5 to 6 percent decline in the quality of new labor-force entrants occurred gradually after 1973. In a simulation model, Martin Neil Baily demonstrated that this could have had only a minor effect on productivity, at least through 1979.[7]

Demographic Change

Another reason cited for a possible drop in the quality of the labor input is that the age-sex mix of the work force has changed substantially. Standard productivity calculations assume that an hour of work is the same regardless of whether it is supplied by an experienced mature adult or an inexperienced teenager. But adult men earn three times as much as teenagers and one and a half times as much as women. Economic theory suggests that these wage differences reflect productivity differences, because if the adult men were not more productive in proportion to their relative wages, their employers would have replaced them with cheaper alternatives.

Adult men made up 58 percent of employed persons in 1957, and this figure had fallen to 45 percent by 1979.[8] Thus if it is true that other demographic groups are only one-third or two-thirds as productive as adult men, this demographic shift should have substantially affected productivity.

Most people are willing to accept the idea that teenagers have low relative productivities, but most people, including the authors, are unwilling to accept the hypothesis that women are inherently less productive than men.[9] However, differences in both wages and productivities need have nothing to do with the intrinsic abilities of the different groups. Women may be confined by custom or discrimination to low-productivity jobs. And, on average, women do have less work experience than men. To see the potential consequences of demographic change, therefore, Baily

7. Martin Neil Baily, "Productivity and the Services of Capital and Labor," *Brookings Papers on Economic Activity, 1:1981,* pp. 1–50. (Hereafter *BPEA.*)

8. Adult males twenty-five years and older. *Employment and Training Report of the President, 1982,* p. 173.

9. The conflicting evidence on male and female wages and their determinants is reviewed in Henry J. Aaron and Cameran M. Lougy, *The Comparable Worth Controversy* (Brookings, 1986).

updated earlier work of George L. Perry and constructed an adjusted labor-input measure in which workers in each demographic group were weighted by the relative wage rate of that group.[10] As the demographic distribution shifted toward lower-wage workers, the adjusted labor-input series declined relative to a conventional measure of hours of work, reflecting a possible decline in skill and experience.

The exercise showed that adjusted labor input grew more slowly than the conventional series over the entire period considered (1950 to 1979), and that the impact of demographic change accelerated after 1968. The changing mix of the work force reduced the effective labor input by 0.2 percent a year before 1968 and by 0.4 to 0.5 percent after 1968. Thus the adjustment does make some difference. Demographic change may be an important reason why productivity slowed after 1968. Since there was no further shift in the trend in 1973, however, the adjustment fails to explain why the slowdown intensified after 1973. Moreover, in the 1980s the demographic change has reversed direction. The labor force is aging as the baby boom generation moves into its thirties and forties.

Work Effort

In a 1979 study, Frank Stafford and Greg J. Duncan analyzed time diaries completed by a sample of several hundred working men and women in the mid-1970s.[11] They found that the average worker spent a considerable amount of time at work but not working. This evidence suggested to some that idleness on the job had caused the slowdown. But because only one year of data was available, the importance of their finding was exaggerated. It is hard to say whether this phenomenon has increased or decreased, let alone whether it accelerated as productivity growth slowed. One suspects that the average worker has always taken a fair amount of slack time.

One way in which work effort might have diminished is through the growth of nonproduction workers as a fraction of the total. Production workers are more closely supervised than nonproduction workers, and their work effort is often determined by the machinery they work with.

10. George L. Perry, "Labor Force Structure, Potential Output, and Productivity," *BPEA, 3:1971*, pp. 533–65, updated in Baily, "Productivity and the Services of Capital and Labor," pp. 8–10.

11. Frank Stafford and Greg J. Duncan, "The Use of Time and Technology by Households in the United States," in Ronald A. Ehrenberg, ed., *Research in Labor Economics*, vol. 3: *1980* (Greenwich, Conn.: JAI Press, 1980), pp. 335–75.

Production workers have less opportunity for leisure on the job. Even if production workers do work harder than nonproduction workers, this fact fails to explain any of the slowdown in productivity growth, for employment of nonproduction workers actually grew a bit more rapidly before 1965 than afterwards.[12]

Conclusions on Labor Quality and Effort

For many people, either a deterioration in work effort or a decline in labor quality plausibly explains weak productivity performance. And that may turn out to be right when all the evidence is in. But to date the evidence for either hypothesis is thin and inconclusive. The changing age-sex mix of the labor force is probably the most important quantitatively, but even this aspect explains only a part of the overall slowdown.

Capital Accumulation and Capital Services

One striking fact about countries with rapid productivity growth compared with countries with slow productivity growth is that the rapid growers invest in new plant and equipment at a much higher rate than do the slow growers. This evidence suggests that inadequate capital formation may have caused the slowdown.

Edward A. Hudson and Dale W. Jorgenson have suggested why capital formation may have slowed, pulling productivity growth down with it.[13] The price of energy began rising in 1970 and jumped sharply in 1973. Hudson and Jorgenson have argued that capital equipment uses energy to operate. Automation can be thought of as a process whereby human power is replaced by machinery that uses energy. A rise in the cost of energy makes automation less attractive and hence discourages capital investment. This hypothesis is plausible and carries added weight because it can be extended to other countries. Energy prices went up for all the industrialized countries.

---

12. If the true labor input equals hours of production workers plus some fraction x, less than unity, times the hours of nonproduction workers, then the growth of the measured labor input is overstated. We tried $\alpha = 0.5$ and found the impact was to make the slowdown even more of a puzzle.

13. Edward A. Hudson and Dale W. Jorgenson, "U.S. Energy Policy and Economic Growth, 1975–2000," *Bell Journal of Economics and Management Science*, vol. 5 (Autumn 1974), p. 461.

Table 2-1. *The Role of Capital Formation in the Slowdown: Nonfarm Business Sector, 1949–86*
Average annual percentage growth

Period	Capital input	Capital per hour
1949–65	2.99	1.97
1965–73	4.54	2.46
1973–79	3.87	1.47
1979–86	3.89	2.36

Source: U.S. Bureau of Labor Statistics, "Multifactor Productivity Measures, 1986," *News*, USDL 87-436, October 13, 1987, table 5.

The first question, however, is whether the pace of capital formation actually fell after 1965 or 1973, or was inadequate to maintain productivity growth. Table 2-1 gives information on capital formation. The growth rate of the capital input was actually higher from 1965 to 1973 than it was before 1965. A stronger case for the inadequacy of investment can be based on the observation that the ratio of capital input to labor input grew less rapidly from 1973 to 1979 than in earlier periods. The 1970s saw a tremendous increase in labor supply in the United States, as young people and women came flooding into the labor force—the demographic shift discussed earlier. To maintain the growth rate of the capital-labor ratio, investment would have had to rise as a share of output, and it did not. Since 1979, however, the capital-labor ratio has grown rapidly again.

The bottom line on whether capital formation is important for the slowdown comes from the comparison of labor productivity and multifactor productivity growth made in chapter 1. As explained, multifactor productivity growth is computed taking into account the effect of changes in the pace of capital formation. Hence a slowing in capital formation cannot be an explanation for a slowing in multifactor productivity growth. At most, it can help explain a slowing in labor productivity growth, and it does not do much of that.

Capital Services

In a study published in 1981 Baily suggested that part or all of the slowdown in productivity growth might be attributable to a deterioration, not in the amount of capital, but rather in the ability of this capital to provide a flow of productive services.[14] If such a decline in capital quality

14. Baily, "Productivity and the Services of Capital and Labor."

had taken place but had not been reflected in conventional capital-input measures, it would have adversely affected both measured labor and multifactor productivity. Baily listed three possible reasons a decline in capital services might have occurred. First, the rise in the cost of energy made some of the existing energy-inefficient capital obsolete. Second, pollution-abatement and worker-safety regulations meant that part of the flow of new equipment failed to add to the ability of capital to produce output. Third, the expansion of foreign trade meant that many factories in the United States became uncompetitive and had to be closed or were underutilized.

Capital equipment is designed to produce a particular product in a particular way. Guesses have to be made about the future cost of the labor and raw materials that a piece of capital will use and about the pattern of demand for the product. If those guesses are wrong, the capital is less productive. Companies with energy-inefficient plants found they were costly to run in 1974. Companies that in the 1960s had invested in plants to produce consumer electronics found that these plants could not produce and sell profitably against foreign competition in the 1970s.

Baily's hypothesis, therefore, is that even though capital formation remained fairly strong after 1965, a lot of the investment decisions turned out to be bad ones. As a result, some capital was scrapped prematurely, and perhaps more important, some capital was never fully utilized.

Is there any evidence to support this view? First, capital productivity has been declining in recent years (as shown in table 1-1). This issue was developed by Barry P. Bosworth for the manufacturing sector.[15] Manufacturers report an estimate of their productive capacity to the Federal Reserve Board. Bosworth found that this capacity in 1980 was down about 24 percent relative to what would have been predicted from the historical relation between capacity and capital investment.[16]

Second, the pattern of the slowdown across industries is consistent with the hypothesis that capital services declined. The manufacturing industries that suffered the worst productivity slowdowns were such capital-intensive industries as chemicals and petroleum refining.[17] A decline in capital ser-

15. Barry P. Bosworth, "Capital Formation and Economic Policy," *BPEA, 2:1982*, pp. 273–317.
16. Ibid., pp. 288–89.
17. Martin Neil Baily, "The Productivity Growth Slowdown by Industry," *BPEA, 2:1982*, pp. 423–54.

vices could be expected to have its greatest effect on capital-intensive industries.

Third, the valuation that investors placed on corporate capital declined dramatically in the 1970s. According to the Council of Economic Advisers, each dollar of corporate fixed investment was valued at about one dollar by stock and bond markets in 1972.[18] By 1979 the value had fallen to 56 cents. The decline means that Wall Street viewed capital as less able to generate productive and profitable capital services.[19]

There is some evidence against the idea. Robert J. Gordon looked in detail at the electric utility industry and concluded that capital obsolescence had not been a major factor.[20] Bosworth, who found the supporting evidence given above, also found that Census Bureau figures on the book value of corporate capital showed no sudden surge in the rate at which capital was scrapped.[21] Charles R. Hulten, James W. Robertson, and Frank C. Wykoff have estimated the effect of the energy crisis on the second-hand prices of various capital assets and found no evidence of accelerated obsolescence.[22] And the market's valuation of capital has risen sharply in the last couple of years, with no evidence of an overall growth recovery.

These objections are not definitive, but their cumulative effect casts doubt on the idea of widespread scrapping of capital. Our own thinking is that capital was used less and rather more inefficiently in the 1970s than in earlier periods, but it probably was not scrapped.

18. *Economic Report of the President, January 1981,* table B-86.
19. Not all this decline was due to a decline in capital services. The discount rate used to value the expected stream of capital earnings probably rose over the period, driven up by high nominal interest rates and an increase in the risk premium. To explain the stock market decline on the basis of interest rates, one has to postulate substantial money illusion because real interest rates were low in the 1970s. Apparently this money illusion failed to affect homeowners, because residential real estate prices rose rapidly over the same period. For a discussion of these issues, see Franco Modigliani and Richard A. Cohn, "Inflation, Rational Valuation and the Market," *Financial Analysts Journal,* vol. 35 (March–April 1979), pp. 24–44; Robert S. Pindyck, "Risk, Inflation, and the Stock Market," *American Economic Review,* vol. 74 (June 1984), pp. 335–51; James M. Poterba and Lawrence H. Summers, "The Persistence of Volatility and Stock Market Fluctuations," Working Paper 1462 (Cambridge, Mass.: National Bureau of Economic Research, September 1984); and Robert J. Shiller, "Stock Prices and Social Dynamics," *BPEA, 2:1984,* pp. 457–98.
20. Robert J. Gordon, "The Productivity Slowdown in the Steam-Electric Generating Industry," Northwestern University, February 1983.
21. Bosworth, "Capital Formation and Economic Policy," p. 287.
22. Charles R. Hulten, James W. Robertson, and Frank C. Wykoff, "Energy, Obsolescence, and the Productivity Slowdown," Working Paper 2404 (Cambridge, Mass.: NBER, October 1987).

Output Measurement and Mix

An enormous variety of goods and services are produced each year. New models of goods are introduced that may be bigger or smaller or different in various characteristics from last year's. Everything from undertakers' services to banking services has to be valued to compute aggregate productivity. Many totally new products and services are introduced each year.

In the United States fairly reliable data are available on private nonfarm business output measured in current dollars. But given the variety of goods and services, it is difficult to form a price deflator that can be used with confidence to divide the increase in the value of business output into a real increase and an inflationary component. Some analysts doubt whether productivity growth is being measured accurately, suspecting that part of the slowdown may just be statistical error.

Problems with Output Measurement

One reason that overall productivity slowed in the mid-1970s is that the level of measured productivity in the construction industry collapsed beginning in 1968.[23] Most observers doubt the validity of the data. There were important technical advances in construction methods and materials after 1968, and independent estimates made of labor requirements for various tasks in the industry showed these requirements continuing to fall after 1968. The problem with measuring productivity for this industry is that each construction project is unique. Coming up with a price index, and hence a value for real output, is almost impossible for a large part of the industry. Steven G. Allen argues that half of the decline in productivity growth is a result of improper deflation.[24] H. Kemble Stokes, Jr., however, takes a more cautious position.[25] Certainly some of the construction problem comes from measurement, but exactly how much is hard to say.

23. Michael F. Mohr estimates that 36 percent of the decline in productivity growth, comparing 1966–73 with 1948–66, is attributable to the collapse in construction productivity. The figure drops to 18 percent of the slowdown for 1973–78. Michael F. Mohr, "Diagnosing the Productivity Problem and Developing an Rx for Improving the Prognosis," Working Paper on Productivity and Economic Growth (Washington, D.C.: Cabinet Council on Economic Affairs, October 1983), table 5.

24. Steven G. Allen, "Why Construction Industry Productivity Is Declining," Working Paper 1555 (Cambridge, Mass.: NBER, February 1985).

25. H. Kemble Stokes, Jr., "An Examination of the Productivity Decline in the Construction Industry," Review of Economics and Statistics, vol. 63 (November 1981), pp. 495–502.

The biggest general criticism leveled at output measurement is that quality change resulting from innovation and better design is not fully captured. Inflation is overstated and real output growth understated. But even if this criticism is correct, it does not necessarily explain the slow-down, because the problem has always existed. Albert Rees, who headed a commission on productivity statistics, has argued that to explain the slow-down in productivity growth in terms of errors in output measurement, one must show that the errors grew worse in the 1970s. That did not happen, says Rees; if anything the Bureau of Labor Statistics (BLS) has been doing a better job of measuring quality change. If so, the "quality-adjusted" slowdown could be even worse than the conventional slowdown.[26]

Others have disagreed, however, arguing that the U.S. economy has become a service economy and that the procedures for estimating price changes really do not work well in the service economy because it is so hard to compare the services provided by a bank or a firm of architects from one year to another.

The Shift to Services

Considering the problem of measurement in the service sector leads to a more general issue: how much have changes in the mix of products and services in the economy damped either actual or measured productivity growth?

Lester C. Thurow has pointed out that average labor productivity is very different in different industries.[27] For example, in 1972 each hour of work in the chemical industry produced $10.02 worth of output, compared with $5.41 for services and $5.87 for trade.[28] The mix of employment in the U.S. economy is shifting away from high-productivity jobs in the indus-trial sector and toward low-productivity jobs in the service sector and retail trade, argues Thurow. If someone who is producing $10 of output per hour is put into a job producing $5 of output an hour, average productivity is bound to decline, he says.

The Thurow argument, which carries immense plausibility and has

26. Albert Rees, "Improving Productivity Measurement," *American Economic Review*, vol. 70 (May 1980, *Papers and Proceedings, 1979*), pp. 340–42.

27. Lester C. Thurow, "Solving the Productivity Problem," in Thurow, Arnold Packer, and Howard J. Samuels, *Strengthening the Economy: Studies in Productivity* (Washington, D.C.: Center for Democratic Policy, 1981), pp. 9–19.

28. Data provided by Ms. Noreen Preston, assistant to Professor Elliott S. Grossman of Pace University.

influenced thinking on industrial policy, is based on a big assumption—
that moving a worker will somehow raise or lower that worker's productivity. In general this proposition is not terribly plausible. For one thing, a
secretary in the chemical industry is probably doing about the same work
as a secretary in the furniture industry. But the issue is deeper than that
because the theory of efficient markets says that moving a worker will
make no difference to overall productivity, even if the specific tasks
change. Changing the industry label on a worker does not in itself have an
impact.

Average labor productivity is high in the chemical business because that
industry uses so much capital, including the technology capital generated
by research and development. Making the economy more capital-intensive
by adding more investment will cause average labor productivity to rise,
but it is adding the capital that does it. Moving workers does not. This is
true whether the extra capital goes into the chemical industry or is used to
make service businesses more capital-intensive (that is, more like the
chemical business).

Qualifications are the bane of economic exposition, but one has to be
made here. If the labor market were perfectly efficient, the industry label a
worker carries would make no difference. But inefficiencies do exist, so
moving workers can in fact raise or lower efficiency and hence can raise or
lower overall productivity. This qualification is minor, though, because
we lack any evidence of trends in efficiency large enough to make a difference.[29]

A similar argument to the one put forward by Thurow has been raised
by William J. Baumol and Edward Wolff.[30] They claim it is not differences in the levels of productivity among industries that are important but
differences in their potentials for growth. As the U.S. economy matures,
more and more of the work force is engaged in service activities, where the
potential for technological change or automation is limited. For example, a
schoolteacher or a waiter or a hotel maid is each doing about the same
things in about the same way as they were done 20 years ago, or even 100
years ago.

Obviously the Baumol and Wolff hypothesis is related to the measurement issues we raised earlier. Baumol and Wolff say that the shift to
services has had an adverse effect on *actual* productivity growth. The
measurement-error view claims that the adverse effect is only on *measured*

29. An exception is agriculture, but this industry is not included here.
30. William J. Baumol and Edward Wolff, ''On the Theory of Productivity and Unbalanced
Growth,'' New York University, November 1979.

productivity growth. Without better measurement techniques it is hard to decide between these two, but there is some evidence that bears on both ideas.

First, the magnitude of the shift to services is often exaggerated. In 1957, manufacturing, construction, and mining (the goods-producing sector) accounted for 40 percent of private nonfarm output. This rate fell to 38 percent by 1973 and to 35 percent by 1979, not a dramatic decline. The relative decline in employment in goods production is greater: from 46 percent in 1947 down to 36 percent in 1979. But the employment share is less important than the output share in productivity calculations.[31]

Second, even if some shift to services has taken place, it is incorrect to say that the whole broadly defined service sector is of uniformly low productivity or unable to achieve productivity growth. The supplying of telephone services, for example, is an industry with a history of above-average productivity growth that was sustained through the 1970s. Even in wholesale and retail trade and personal and business services, output per hour grew at about the same rate between 1957 and 1973 as it did in manufacturing.

In a 1982 paper Baily calculated how much of the decline in productivity growth had occurred within the major industries of the economy and how much was the result of shifts in industry mix.[32] He found that little of the slowdown could be accounted for by mix effects. Independent work by Frank M. Gollop agrees with this result.[33] It is a result that surprises many people and occurs mostly because the change in the mix of production has not been consistently toward sectors with low growth potential or low measured productivity growth. Parts of the service-producing sector have had good productivity growth (communications), while parts of the goods-producing sector (construction and mining) have done poorly.

The Computation of Output

The way analysts compute an industry's share of output—whether in constant dollars or current dollars—also affects our view of the productivity slowdown. In computing the growth of real output, the standard

31. Think of a two-sector economy. One sector consists of personal services with no productivity growth. The second sector consists of a highly automated sector with only a tiny fraction of employment. If output shares remain at 50–50 for the two sectors, productivity growth will remain at about half the rate of the growth in the automated sector.

32. Baily, "Productivity Growth Slowdown," pp. 445–54.

33. Frank M. Gollop, "Analysis of the Productivity Slowdown: Evidence for a Sector-Biased or Sector-Neutral Industrial Strategy," in William J. Baumol and Kenneth McLennan, *Productivity Growth and U.S. Competitiveness* (Oxford University Press, 1985), pp. 160–86.

method is to measure output shares in constant rather than current dollars. For industries whose prices are falling relative to the average of other industries—for example, the computer business—the current-dollar share of output will fall relative to the constant-dollar share. Since the current-dollar shares would provide a better way to compute output growth, traditional methods introduce a measurement error, which can be quite large in industries with falling relative prices.

Formally, the nominal value of output in a given year (V) is the total of the nominal values of output in all the constituent industries in the economy.

(2-1) $$V = \Sigma_i V_i \ (i = 1, \ . \ . \ . \ \text{industries}).$$

The real output of industry i (Q_i) is computed using a price index (P_i) that estimates for a given year how much the industry's prices have increased or decreased relative to a fixed base year.

(2-2) $$Q_i = V_i/P_i.$$

Aggregate real output is then the sum of the industry real outputs.

(2-3) $$Q = \Sigma Q_i.$$

The unit of measurement is dollars of the base year. The rates of growth of current- and constant-dollar output are given by

(2-4) $$\frac{\Delta V}{V} = \sum_i \alpha_i \frac{\Delta V_i}{V_i}, \qquad \frac{\Delta Q}{Q} = \sum_i \beta_i \frac{\Delta Q_i}{Q_i}.$$

The growth rates in the aggregates are weighted averages of the growth rates by industry. The α's are the shares of the industries in nominal output and the β's are the shares of the industries in real-dollar or constant-dollar output. The relationship between the α's and β's is as follows:

(2-5) $$\alpha/\beta = P_i/P,$$

where $P = V/Q$. It is the implicit deflator for aggregate output. This relation shows that the current-dollar share of output will fall relative to the constant-dollar share for industries whose prices are falling relative to the average of other industries. This is the source of measurement bias in the calculation of constant-dollar output.

The pattern in the United States is that rapidly growing industries such as telephonic communications and air transport services have falling relative prices. Technological change leads to cost reductions that then stimulate demand. This means, in practice, that for years following the fixed base year, the constant-dollar shares of the rapidly growing industries are rising relative to their current-dollar shares.

Production theory and indeed common sense suggest that using the current-dollar output shares would be a more appropriate way to compute the rate of growth of real output, rather than the constant-dollar shares implied by the standard procedures.[34] If society allocates 10 percent of its current income to the output of an industry, then that is the weight it should have in the aggregate growth rate, not the share that results from extrapolating constant-dollar shares from an arbitrary base year. Contemporary procedures overstate real output and productivity growth for years following the base year and understate growth for years preceding the base year.

The Commerce Department used 1972 as a base year until fairly recently, so the slowdown in productivity growth after 1972 was being understated, particularly in manufacturing, where the relative price changes are quite pronounced. When the base year was changed to 1982, this change by itself resulted in a larger slowdown. Yet the total of all revisions made at that time failed to change the size of the slowdown much because at the same time the base year was being changed, the new price index for computers was introduced. This change in methods boosted measured real output growth. The two changes offset each other. In the future, however, there will not be an offset. Computer prices have been falling at about 20 percent a year relative to the average for manufacturing. The constant-dollar share of computers in manufacturing output is now small for the years before 1982, and will grow larger and larger following 1982. This adds to the post-1982 recovery of productivity growth in manufacturing in a way that is partly a statistical illusion.

Management Failures

In 1980 Robert H. Hayes and William J. Abernathy published an enormously influential article proposing that slow productivity growth and an inability of U.S. businesses to compete overseas were the results of serious failures by top management.[35] The authors argued that American manag-

---

34. For a discussion of this issue, see Franklin M. Fisher and Karl Shell, *The Economic Theory of Price Indices: Two Essays on the Effects of Taste, Quality, and Technological Change* (Academic Press, 1972).

35. Robert H. Hayes and William J. Abernathy, "Managing Our Way to Economic Decline," *Harvard Business Review*, vol. 58 (July–August 1980), pp. 67–77.

ers had abandoned long-term technological superiority as a strategy for success and instead geared their decisionmaking to short-term profits.

Modern management techniques are blamed for this problem. Scientific management emphasizes "analytic detachment rather than the insight that comes from 'hands-on' experience and . . . short-term cost reduction rather than long-term development of technological competitiveness."[36] Moreover, the rise of MBA graduates with their kits of management tools has promoted the idea of an "interchangeable manager," someone who can come into a company and run it without detailed knowledge of the industry or its technology. This system encourages managers to stress such things as takeovers, financial manipulation, and the setting up of profit centers to achieve quick results before they move on to the next industry.

The ideas expressed by Hayes and Abernathy struck a responsive chord, especially among scientific and engineering staffs in U.S. corporations. The R&D people whom we interviewed in our work on innovation frequently spoke of the frustrations they had felt in the 1970s in persuading senior management to support new product and process development. But for all its popularity the view that the productivity growth slowdown was caused by an exogenous wave of poor management is questionable.

The fact that all the major industrialized countries experienced a slowdown at about the same time has to be more than a coincidence. Yet the idea of managers in New York, London, Bonn, and Paris simultaneously and abruptly becoming incompetent after 1973 seems pretty farfetched. Indeed the critics of U.S. managers usually compare them unfavorably with foreign managers, notably the Japanese, who allegedly have a much longer time horizon. But Japan also had a sharp slowdown in its productivity growth.

The case for poor management is presented without serious evidence to support it. It is easy enough to find examples that illustrate poor management decisions, especially with the advantage of hindsight. But the advocates of the hypothesis have not yet done the hard work needed to demonstrate that a major changing of the guard really took place and that scientific managers have actually performed worse than those relying only on experience.

Rather than postulating a sudden attack of poor management, it is far more plausible to link management failure to other explanations of the slowdown. The energy crisis, inflation, fierce foreign competition, and the

36. Ibid., p. 68.

drive for environmental regulation challenged managers to deal with these problems without sacrificing productivity. They did not succeed. Office automation provided the means for reducing white-collar costs, but too often the potential gains have not been realized. Capital was less productive in the 1970s than earlier, and management was responsible for the investment decisions and the running of the plants. Innovation slowed in some industries, and while depletion of opportunities was important, courageous management might have spurred the search for new areas of opportunity. Perhaps old-fashioned managers would have dealt with these opportunities and problems more effectively than the new breed of business-school graduates, but we suspect that both old-style and new-style managers made their share of mistakes.

Government Regulatory and Demand Policies

Without doubt certain government policies have adversely affected productivity growth during the past fifteen years. That does not necessarily mean that such actions were wrongly taken. For example, the slow growth of aggregate demand and the deep recessions of 1975 and 1982 were undertaken to fight inflation. The temporary or even permanent loss of productivity that resulted was part of the price. Similarly, regulating pollutants and promoting worker safety are policies undertaken for sound reasons, even though critics may question their cost-effectiveness. Without drawing any conclusion about the overall wisdom of government actions, we can try to assess how they have influenced productivity.

Health and Safety Regulation

The negative impact of Environmental Protection Agency (EPA) and Occupational Safety and Health Administration (OSHA) regulations on productivity has been studied by a number of economists.[37] Most have found the impact of regulation was pretty small.

37. Denison, *Trends in American Economic Growth;* J. R. Norsworthy, Michael J. Harper, and Kent Kunze, "The Slowdown in Productivity Growth: Analysis of Some Contributing Factors," *BPEA, 2:1979,* pp. 387–421; Gregory B. Christainsen and Robert H. Haveman, "Public Regulations and the Slowdown in Productivity Growth," *American Economic Review,* vol. 71 (May 1981, *Papers and Proceedings, 1980*), pp. 320–25; Robert W. Crandall, "Pollution Controls and Productivity Growth in Basic Industries," in Thomas G. Cowing and Rodney E. Stevenson, eds., *Productivity Measurement in Regulated Industries* (Academic Press, 1981), pp. 347–68; and Wayne B. Gray, "The Impact of OSHA and EPA Regulation on Productivity," Working Paper 1405 (Cambridge, Mass.: NBER, 1984).

Edward Denison's study, a particularly careful one, finds that pollution abatement and worker safety and health subtracted only 0.12 percentage point from the 1973–82 growth rate, compared with 1948–73. The bulk of this cost (0.10 percentage point) comes from his estimates of the requirements to meet pollution abatement standards. OSHA regulations had a small effect, except in the mining industry, argues Denison, because they largely codified existing rules. In many cases industry organizations supported the rules because most firms were already observing them.[38]

There is one important exception to the studies concluding that the impact of regulation was minor. Wayne B. Gray finds a strong cross-sectional correlation between industries with large slowdowns in productivity growth and those that had been heavily affected by regulation. He estimates that 35 to 39 percent of the decline in total factor productivity growth in his sample of 450 manufacturing industries was the result of EPA and OSHA regulation.[39] This figure is three to four times as high as Denison's.

The Gray study should be taken seriously because he has constructed a valuable data series that correlates the degree of regulatory impact with the slowdown at a level of industry detail that is not available elsewhere. The reasons for caution in reading the Gray results are, first, they are outliers compared with what others found, and second, his econometric model really emphasizes regulation. He does not have many alternative explanations of the slowdown in his data, and the regulation variables are much more extensive than any other variables.

On balance, we estimate something between the Denison and Gray figures. Denison is conservative because he does not measure any impact of regulation that he cannot quantify fairly directly. Gray is likely to attribute to regulation some of the impact of other factors. A range of 0.20 to 0.25 percent a year is a reasonable figure on the decline in productivity growth in the nonfarm economy from 1970 to 1980 resulting from EPA and OSHA regulations. Since 1980 the effect has been small because regulation has become no more stringent.

Inflation

The burst of inflation that occurred in 1973 was followed by disastrous productivity figures in 1974 and 1975. The acceleration of inflation after

38. Denison, *Trends in American Economic Growth*, pp. 37, 66–69.
39. Gray, "Impact of OSHA and EPA Regulation," pp. 27, 35.

1978 was again followed by a period of poor productivity performance. These experiences have prompted analysts such as Peter K. Clark and Michael Mohr to argue that inflation disrupted economic efficiency and caused at least part of the slowdown.[40] But we have never found this argument persuasive because no one has shown how the loss of output could have been large enough to be significant. Clark argues that inflation results in relative price errors and that these errors cause labor to be misallocated. But the efficiency loss equals the product of the gap between actual and efficient marginal productivity and the likely magnitude of the extent of the error in labor allocation. The result in Clark's model looks like a second-order small number. Mohr argues that inflation led management in uncompetitive industries (he cites steel and autos) to focus on raising prices rather than cutting costs. There may be some truth to this argument, but again we think it more plausible that inflation was just one of several problems that diverted managerial attention away from productivity.

Clark himself has pointed out that the inflation hypothesis has been weakened by the continuation of the slowdown despite the absence of serious inflation in the past few years.[41]

Slow and Variable Demand Growth

Output growth correlates strongly with productivity growth, both over time within countries and across countries. This correlation may mean that strong growth in product demand stimulates productivity growth. Or it may mean that rapid productivity growth allows or promotes rapid growth in output and demand. The conventional view holds both interpretations are correct, because demand affects productivity over short periods, such as the business cycle, while productivity increases have the reverse effect of allowing growth in trend demand and output over the long term.

This conventional dichotomy between short- and long-term effects would break down if aggregate demand were chronically low, or chronically variable, and if either of these discouraged innovation, risk-taking, and capital formation. In these circumstances, weak demand or frequent recessions could reduce trend productivity growth. An obvious test case

40. Peter K. Clark, "Inflation and the Productivity Decline," *American Economic Review,* vol. 72 (May 1982, *Papers and Proceedings, 1981*), pp. 149–54; and Mohr, "Diagnosing the Productivity Problem," pp. 29–31.

41. Peter K. Clark, "Productivity and Profits in the 1980s: Are They Really Improving?" *BPEA, 1:1984,* pp. 133–67.

Table 2-2. *Causes of the Productivity Growth Slowdown according to Denison and Mohr*

Study and determinant	Effect on post-1973 growth (percentage points)	Percentage contribution to slowdown
Denison	Slowdown in growth of output per hour, nonresidential business, 1973–82, compared with 1948–73	
Education	0.10	−4.1
Weather in farming	0.03	−1.2
Age-sex composition	−0.02	0.8
Inventories	−0.09	3.7
Nonresidential fixed capital	−0.06	2.5
Land	−0.02	0.8
Reallocation from farming	−0.19	7.8
Reallocation from nonfarm self-employment	−0.17	7.0
Pollution abatement	−0.10	4.1
Worker safety and health	−0.02	0.8
Dishonesty and crime	−0.05	2.0
Economies of scale	−0.15	6.1
Intensity of demand	−0.23	9.4
Residual productivity[a]	−1.47	60.2
Total slowdown	−2.44	100.0
Mohr	Slowdown in growth of output per hour, private business, 1973–78, compared with 1948–66	
Changes in the composition of output	−0.2 to −0.4	10 to 20
Effect of recessions, strikes, weather, energy on output	−0.6 to −0.8	30 to 40
Capital formation and decline in capital quality	−0.5 to −0.7	25 to 35
Age-sex composition	−0.1 to −0.2	4 to 10
Federal regulations	−0.2	8 to 10
Residual productivity[a]	−0.4 to −1.0	20 to 50
Total slowdown	−1.94	100

Sources: Top panel from Edward F. Denison, *Trends in American Economic Growth, 1929–1982* (Brookings, 1985), p. 37; bottom panel based on Michael F. Mohr, "Diagnosing the Productivity Problem and Developing an Rx for Improving the Prognosis," Working Paper on Productivity and Economic Growth (Washington, D.C.: Cabinet Council for Economic Affairs, October 1983), pp. 10–11.

a. Advances in knowledge and miscellaneous determinants.

for this hypothesis is the Great Depression. Productivity fell dramatically between 1928 and 1933. It then recovered rapidly, but did it go back to the usual trend? Denison calculates that national income per workhour grew 2.0 percent a year from 1929 to 1941. This figure is a little below the long-run trend, but not much. It suggests that the long-run effect of the Depression was slight. However, Denison's productivity growth residual (advances in knowledge and miscellaneous determinants) grew very slowly over the same period.[42] And this finding *is* consistent with the view that innovation was impeded. The reason for the difference in conclusions is that Denison attributes much of the increase in output per hour leading up to 1941 to the rapid growth in demand in 1940 and 1941.

We are not sure Denison is correct in this. We have found that separating out the impact of trends and cycles in productivity data is one of the most difficult tasks there is. Since it is almost impossible to be confident about the very short-run effects of demand fluctuations on year-to-year productivity changes, it is extremely hard to determine if weak or variable demand has had a longer-run impact on the productivity trend. The lesson of the Great Depression is more unclear than one would like. And the conclusions from postwar cycles are even harder to be certain about. The effect of demand on growth is not something we feel is well understood, and we will return to this issue in our industry studies.

Contributors to the Slowdown Except Innovation

Denison and Mohr have reviewed the causes of the slowdown in productivity growth and provided quantitative estimates of the contributing factors. Table 2-2 gives their results. Denison's figures are all based on things that he can measure; he does not make quantitative guesses. He finds that the slowdown is partly the result of a hundred small wounds, but mostly remains a mystery. Sixty percent of the slowdown after 1973 is attributed to things that are not specifically identified and are therefore attributed by default to a decline in the "advances in knowledge" residual item. Mohr is more heuristic from the beginning. He gives ranges of effects for broad groups of contributory factors. In general, he attributes much more to the specific causes than does Denison. He also argues that the rise in energy prices and the decline in capital quality or capital services

---

42. Denison, *Trends in American Economic Growth*, pp. 76, 98.

have made serious contributions to the slowdown, whereas Denison does not count these in.

Denison's methodology is a good one, but it is intrinsically conservative. It seems likely that regulation and labor quality have had bigger effects than his table indicates. Mohr, on the other hand, is a bit heroic in some of the numbers he derives. We fail to see a clear basis for attributing a 1.0 to 1.5 percentage point decline in growth to recessions, energy, weather, and capital services.

Looking at studies such as those by Denison and Mohr and at our own evaluation, we can draw several conclusions about the decline. The slowdown in productivity growth before 1973 was smaller than the one since 1973 has been, was unique to the United States, and can be associated with factors that can be explained, or at least identified. These include the collapse of productivity in mining and construction after 1968, the entry of less-experienced workers into the labor force, and the reduction in the intensity of factor use as the 1960s boom was restrained.

There are some clear villains for the post-1973 slowdown also. The ones described above continued after 1973, and such factors as the energy crisis, inflation, regulation, variable demand, and capital quality became important. But after making plausible estimates for these, there remains a post-1973 decline in the productivity residual of about 1.0 to 1.25 percent a year. There is disagreement about the importance of the individual contributors to the slowdown, but a consensus exists that an important residual slowdown remains once all of them have been taken into account. *The unexplained slowdown after 1973 is the basic puzzle to be addressed in the remainder of this book.*

As we examine this puzzle it is worth clarifying terms. "Technological progress" and "technical progress" are used interchangeably to indicate increases in multifactor productivity associated with increased knowledge. "Invention" refers to an idea for a new product or process, and "innovation" refers to the results of an earlier invention at a point where the new product or process is introduced or used in production. "Technological opportunities" indicate areas where there is scope for inventions and innovations.

R&D and Innovation

Might the unexplained decline in growth be the result of a collapse in the pace of technological change? There are three broad possibilities that would explain such a collapse. The first is that the available technological opportunities—those for which the basic science and perhaps directions for research were known—were temporarily or permanently depleted. The second is that the opportunities were there, but managers failed to exploit them. The third is that innovations were made, but they did not result in productivity increases.

Were there technological opportunities? To some economists the idea of periods of slow technological advance seems a natural one. Joseph A. Schumpeter, a leading American economist of the interwar years, made spurts or waves of innovation central to his theories of economic development and the business cycle.[43] Supporters of this view point to the past fifty years as an example. The period from the thirties until the end of World War II may have experienced slow growth, they say, partly because the Great Depression and the war depressed the ability of the economy to exploit scientific advances fully. Once the war was over there was a spurt of growth that lasted until 1965, then growth fell back to more normal levels.[44] But if there are many economists who find it plausible that technology has slowed, there are plenty of economists and scientists who find this idea highly implausible. In the midst of the electronics revolution it is shocking to suggest that innovation may have slowed. It is an idea that needs careful review, and the place to start is with research and development spending. Economists, at least, have linked the pace of technological change to the magnitude of the resources devoted to R&D and the payoff from R&D.

The Role of R&D

To an important extent innovation grows out of advances in basic scientific knowledge that are then applied by companies to their products and processes. And a key mechanism by which technical change is brought into production is R&D performed in industry. Companies draw on scientific advances made in universities and research labs, and perform applied

43. Joseph A. Schumpeter, *The Theory of Economic Development: An Inquiry into Profits, Capital, Credit, Interest, and the Business Cycle*, trans. Redvers Opie (Harvard University Press, 1934).

44. William Baumol has argued this view, concluding that the recent slowdown in productivity growth is not something to worry about and that the economy is simply returning to its long-run growth trend. That is going too far. We argued in chapter 1 that the slowdown is a persistent and serious problem. Since 1973 the economy has been well below its long-run trend of growth. See William J. Baumol, "Productivity Growth, Convergence, and Welfare: What the Long-Run Data Show," *American Economic Review*, vol. 76 (December 1986), pp. 1072–85.

research and development to turn these advances into marketable products or new processes. Not all innovation, however, uses new basic knowledge. Company labs may simply come up with new ways of using existing scientific knowledge.[45] In either case, a growth in R&D seems necessary to maintain the pace of innovation. Thus if in the 1970s industry devoted inadequate resources to R&D, this might have led to weakness in the level of innovation. Is that view correct?

Define R as the input of R&D into production, ignoring for the present how R is defined or computed. Then output, Q, is assumed to be a simple (Cobb-Douglas) function of the capital input, K, the labor input, L, and the R&D input, R.

$$(2\text{-}6) \qquad Q = AK^{\alpha}L^{\beta}R^{\gamma}.$$

In this specification, α, β, and γ are fixed parameters; they are the elasticities of output with respect to the three inputs. The intercept term is A, and it may increase over time as a result of technical change or organizational innovations unrelated to R&D.

An increase in the R&D input, according to equation 2-6, increases the amount of output produced with a given quantity of capital and labor. In other words, *an increase in the R&D input increases multifactor productivity*. The magnitude of the contribution of R&D to multifactor productivity growth also follows from equation 2-6:

$(2\text{-}7)$ $\quad$ Contribution of R&D to = γ times the rate of growth of MFP growth $\qquad$ the R&D input

$$= \gamma\,(d\ln R/dt).$$

The commonest approach to defining the concept of the R&D input, R, is that it is a stock, computed in the same way that the stock of capital is computed. The stock of R&D consists of a weighted average of R&D spending over a number of years. Interviews with R&D directors and widespread empirical evidence suggest that there is a lag of at least a couple of years from the time of a given R&D expenditure until the results show up in production. Sometimes the lag may be as much as fifteen years. Superimposed upon this lag structure is a depreciation effect. The result of these two forces is that the appropriate weights to apply to current and past

45. An analysis of the innovation process and of the complex links between R&D and the stock of basic knowledge is given in Stephen J. Kline, "Research, Invention, Innovation, and Production: Models and Reality," Report INN-1 (Stanford University, Department of Mechanical Engineering, February 1985).

R&D spending form an inverted U, perhaps with a tail extending into the past. Not everyone, however, uses such a procedure. In fact, some studies seem to find the strongest impact on output from current or very recent R&D. This finding suggests that either the lags are very short, or else the correlation is not indicative of causality. A good year for product demand or some unique opportunity that presents itself may lead to high levels of both productivity and R&D spending. This means that there is no consensus on how to compute R, and there are question marks about some estimates of the appropriate lag and the appropriate depreciation rate to use in such a computation.

One common way to finesse the problem of computing R, and also a way to avoid making a direct estimate of γ, is to assume that the rate of depreciation of R&D is small. The assumption is a reasonable one if new methods and ideas build on and add to old knowledge and methods. It implies the following:

(2-8)   Rate of growth of R&D $= d\ln R/dt = S/R,$

where S is annual spending on R&D, perhaps lagged a couple of years. The rate of return to R&D, ρ, is equal to the marginal product of R, which means that equation 2-6 gives the following:

(2-9) $\rho = \gamma Q/R.$

So equations 2-8 and 2-9 together give the result used by Terleckyj:[46]

(2-10)   Contribution of R&D $= \rho S/Q$
              to MFP growth

                              $=$ rate of return to R&D times
                                  the ratio of R&D spending
                                  to output.

To go from equation 2-10 to a calculation of the effect of R&D, the estimate of the rate of return must take into account that R&D spending consists of outlays on capital and labor that, in practice, are included in K and L. Roughly speaking, R&D contributes to MFP to the extent that its rate of return exceeds the return on conventional capital. An estimate is needed of the rate of return to R&D and such an estimate should look at the gap between social and private returns to R&D.

---

46. See Nestor E. Terleckyj, *Effects of R&D on the Productivity Growth of Industries: An Exploratory Study* (Washington, D.C.: National Planning Association, 1974).

The Private and Social Returns to R&D

Part of the knowledge R&D generates provides benefits not to the company performing the R&D, but to other companies and to consumers. Only part of the *social* return to R&D becomes a *private* return to the company that has paid the bill (the "appropriability problem"). There are several ways in which the spillovers from R&D occur. Competitors may copy or "reverse engineer" new products. They may hire away key personnel. And sometimes, as in the electronics industry, key personnel become competitors by setting up their own companies.

The situation is not all bad. Consumers benefit when information about new technology spreads to several companies and this generates competition. A system that allows the innovating company to be a monopolist is not the best one. To estimate the contribution of R&D to growth, the analyst must look at the social rate of return.

Edwin Mansfield has been a leader in investigating the social and private returns to R&D. In a recent paper, he summarized his own work and that of others.[47] His principal study of a group of specific innovations found a social rate of return of 56 percent and a private rate of 25 percent. He also looked at the private return to one of America's largest companies and found a private rate of return of 19 percent and a social rate of return *at least* twice the private rate.

A study by Robert R. Nathan Associates and one by Foster Associates used Mansfield's methodology on different data.[48] Nathan Associates found a social rate of return of 70 percent versus a private rate of 36 percent. Foster Associates found a social return of 99 percent versus a private rate of 24 percent, only a quarter of the social rate.

Mansfield has also analyzed the patent system and why it does not solve the appropriability problem. Patents cannot prevent imitation, although they raise the cost of it a little. Mansfield reports that within four years of their introduction, 60 percent of the successful patented innovations have been copied. Outside of the ethical drug industry, patenting raises the cost of imitation by less than 10 percent. In the drug industry, the cost is raised

47. Edwin Mansfield, "Microeconomics of Technological Innovation," in Ralph Landau and Nathan Rosenberg, eds., *The Positive Sum Strategy: Harnessing Technology for Economic Growth* (Washington, D.C.: National Academy Press, 1986), pp. 307–25.

48. Robert R. Nathan Associates, *Net Rates of Return on Innovations*, 2 vols., report prepared for the National Science Foundation (Washington, D.C., 1978); and Foster Associates, *A Survey on the Net Rates of Return on Innovations*, 3 vols., report prepared for the National Science Foundation (Washington, D.C., 1978).

by 30 percent. Patents do not enable innovating companies to appropriate returns fully because other companies innovate around the patents.[49]

The conclusion of Mansfield's line of research, therefore, is that a substantial gap exists between the private and social rates of return despite the availability of patents. The social rate of return is between 50 and 100 percent, so to be conservative we will say that the excess return to R&D is 35 to 60 percent above the return to ordinary capital.

The Contribution of R&D to Growth and the Slowdown

Industry-funded R&D varied between about 1.0 and 1.5 percent of private nonfarm GNP during the 1960s. With the excess rate of return between 35 and 60 percent, equation 2-10 predicts that this flow of R&D would result in multifactor productivity growth of 0.35 and 0.90 percent a year. Since the trend of multifactor productivity growth at the time was about 1.75 percent a year, industrial R&D contributed from one-fifth to one-half of all multifactor productivity growth.

The U.S. government also funds R&D, most of it in defense. There are spillovers from this federal R&D, but their size is not well known, nor are clear estimates available of the direct impact of this R&D on productivity.[50] As a best guess, we will assume that each dollar of federal funds makes about the same contribution to productivity as 25 cents of industry funds. Federally funded R&D was about 2.0 to 2.2 percent of private nonfarm GNP in the 1960s, equivalent, by our 25-cent rule, to 0.5 percent in industry funds. This means that an additional 0.175 to 0.30 of multifactor productivity growth was contributed by federal R&D.

The bottom line, therefore, is that R&D spending by industry and the federal government together contributed between 0.53 and 1.20 percent a year to multifactor productivity growth in the 1960s. This represents between 30 and 69 percent of the total.

49. Moreover, tightening the patent system might fail to solve the problem while restricting the diffusion of new technology, to the detriment of consumers. The effort of innovating around a rival's patent imposes private and social costs on the industry. And costly strategic behavior might be fostered by tight patent restrictions—for example, patent applications may be submitted too hurriedly in order to gain strategic advantage. See Morton I. Kamien and Nancy L. Schwartz, "Market Structure and Innovation: A Survey," *Journal of Economic Literature*, vol. 13 (March 1975), pp. 1–37.

50. For a review of this issue and other citations, see Harvey Brooks, "National Science Policy and Technological Innovation," in Landau and Rosenberg, eds., *Positive Sum Strategy*, pp. 119–67.

The rate of growth of R&D spending slowed in the 1970s. While industry-funded R&D, after adjusting for inflation, had climbed 6.2 percent a year from 1960 to 1969, this figure fell to 2.4 percent a year from 1969 to 1977.[51] The timing of this slowdown does not coincide with the main decline in productivity growth. But the discrepancy can be accounted for if there is a several-year lag between R&D spending and observed productivity growth. Thus a slowdown in industrial R&D starting in 1969 could well have been expected to show an effect on productivity around 1973.

Federally funded R&D followed a similar course. Although the growth in federal R&D spending over the entire period 1960–77 was much less than the growth in industry funding, the federal commitment showed a weakening of spending in the 1970s almost in parallel with private funding. The peak of federal spending was in 1970: from 1960 to 1970, real federal R&D spending rose 5.9 percent a year. It then not only stopped growing but actually fell almost 5 percent from 1970 to 1977.

Could these growth declines have accounted for the observed decline in multifactor productivity growth? Given an equivalency rule of 25 cents on the dollar for federal R&D, the answer is clearly no, or not much of it anyway. The contribution of R&D to growth depends on the ratio of spending to output. And output also grew more slowly in the 1970s, so the ratio declined relatively little for industry-funded R&D. Federally funded R&D had its growth cut back more, but its impact on multifactor productivity is smaller, by assumption. Using the same approach described earlier, we can explain at most a decline of 0.15 percent a year in multifactor productivity growth as a result of weakness in R&D spending. Even this figure is stretching it; 0.1 percent is probably closer to the mark.

Other Estimates of the Effect of the Slowdown in R&D Spending

One economist, John W. Kendrick, challenges the limited role we have assigned to federal R&D. He believes that the declines in both industry-funded and federally funded R&D were important causes of the productivity growth slowdown. He computes an effective R&D stock, and finds that after 1973 the decline in R&D growth knocked 0.5 percentage point a

---

51. Data on R&D spending from U.S. National Science Board, *Science Indicators: The 1985 Report* (Washington, D.C.: National Science Foundation, 1985), p. 218. Growth rate calculated using natural logs.

year off the productivity growth rate.[52] This is much bigger than the effect we found.

Bosworth has criticized Kendrick on two grounds.[53] One is that Kendrick gives federally funded R&D equal status with industry-funded R&D as a contributor to productivity growth. That is, Kendrick blames much of the slowdown in multifactor productivity growth on the decline in federally funded R&D after 1970. We agree with Bosworth's criticism. While our 25-cent rule is arbitrary, giving equal weight to both types of funding seems unrealistic. Most federally funded R&D goes for defense, space, and health. In 1971 these three accounted for 80 percent of the total, and most of this R&D has no *direct* effect on measured productivity.[54] Improvements in weaponry, expeditions to space, and enhancement of health care may promote our national well-being, but they are not counted directly in productivity measures, because these are either government activities or are unmeasured quality changes.[55] Kendrick points to the spillover benefits from defense and space research, but claiming that the decline in such research in the 1970s played much of a role in the productivity slowdown is implausible. The spillovers just are not large enough or quick enough.

The second criticism of Kendrick is that in computing his effective R&D stock, he assumes the return to R&D declined. The objection here is not that this assumption is necessarily wrong, but that it confuses two issues. If the pace of technological advance has slowed, we want to know to what extent this is because the economy put fewer resources into R&D, and to what extent the payoff from a given quantity of R&D declined. These issues need to be separated because they have very different implications. Suppose changes in taxes or the interest rate have reduced the quantity of R&D, and this reduction in turn has retarded growth. Then the

52. John W. Kendrick, "The Implications of Growth Accounting Models," in Charles R. Hulten and Isabel V. Sawhill, eds., *The Legacy of Reaganomics: Prospects for Long-Term Growth* (Washington, D.C.: Urban Institute Press, 1984), p. 28. See also Kendrick, "Productivity Trends and the Recent Slowdown: Historical Perspective, Causal Factors, and Policy Options," in William Fellner, ed., *Contemporary Economic Problems* (Washington, D.C.: American Enterprise Institute, 1979), pp. 17–69.

53. Barry P. Bosworth, *Tax Incentives and Economic Growth* (Brookings, 1984), pp. 32–33.

54. National Science Board, *Science Indicators*, p. 227.

55. Health care can improve the productivity of the work force, but the evidence of improved capacity to work is pretty mixed. See Martin Neil Baily, "Aging and the Ability to Work: Policy Issues and Recent Trends," in Gary Burtless, ed., *Work, Health, and Income among the Elderly* (Brookings, 1987).

obvious solution will be to commit more resources to R&D and restore growth. If, however, the root of the problem is reduced opportunities for innovation, then programs to boost R&D will have a lesser effect.

Kendrick aside, the basic consensus among economists agrees with our calculation. The decline in the *quantity* of R&D performed in the 1970s was not in itself large enough to take more than 0.1 or possibly 0.2 percentage point a year off the productivity growth rate.[56] If the pace of innovation has slowed, it is not primarily because of the reduction in the growth of R&D spending.[57]

Econometric Evidence on the Return to R&D

Over the years numerous econometric estimates have been made of the elasticity of output with respect to R&D (α) and the rate of return to R&D (ρ).[58] And there are several recent studies of whether either or both of these parameters declined in the 1970s.

In 1980 Zvi Griliches surveyed the results of several studies of his own and by others and concluded: "If these findings are to be taken at their face value, they imply [a] . . . collapse in the productivity of R&D."[59] In later studies, however, particularly one published in 1986, Griliches has come to dramatically different conclusions.[60] He found that the contribution of R&D to productivity "has not declined significantly in recent years, in spite of the overall slowdown in productivity growth and the general worry about a possible exhaustion of technological opportunities."[61] This same finding was suggested earlier by F. M. Scherer and has been replicated since, using a more complete data set, by Frank R. Lichtenberg and Donald Siegel.[62]

56. See Zvi Griliches, "R&D and the Productivity Slowdown," *American Economic Review*, vol. 70 (May 1980, *Papers and Proceedings, 1979*), pp. 343–48.

57. There is another reason for skepticism about the importance of R&D to the slowdown. Japan and Europe had productivity slowdowns, but sustained the rate of growth of R&D relative to output.

58. The literature through 1971 is reviewed and summarized in National Science Foundation, *Research and Development and Economic Growth/Productivity* (Washington, D.C.: NSF, 1972).

59. Griliches, "R&D and the Productivity Slowdown," p. 346.

60. Zvi Griliches and Frank Lichtenberg, "R&D and Productivity Growth at the Industry Level: Is There Still a Relationship?" in Griliches, ed., *R&D, Patents, and Productivity* (University of Chicago Press, 1984), pp. 465–96; and Griliches, "Productivity, R&D, and Basic Research at the Firm Level in the 1970's," *American Economic Review*, vol. 76 (March 1986), pp. 141–54.

61. Griliches, "Productivity, R&D, and Basic Research," p. 152.

62. F. M. Scherer, "R&D and Declining Productivity Growth," *American Economic*

As is the case in all econometric studies, different sets of data or methods give different results. And there are criticisms one can make of all the studies. But the accumulation of evidence from more recent work is that the correlation between R&D and multifactor productivity remained strong in the 1970s. What does this mean?

An important argument raised by Scherer is as follows.[63] He points out that several of the econometric estimates are based upon equation 2-10 and they show that it is the rate of return to R&D that has remained high. He argues that if companies responded to shrinking technological opportunities by cutting back the quantity of R&D, this would preserve the rate of return. The continued strong correlation between R&D and multifactor productivity, according to Scherer, fails to prove opportunities stayed strong.

Scherer's point is true enough, but it does not solve the underlying dilemma. The fact is companies did not make the kind of large cutbacks in the quantity of R&D that Scherer's story demands. Equation 2-10 tells us that the contribution of R&D to growth is the product of the rate of return to R&D times R&D intensity. Since in the 1970s the rate of return was unchanged (according to the econometric estimates) and since R&D intensity remained more or less at normal levels, it appears that R&D was boosting productivity growth, even though the Denison-style technical change residual was zero or even negative.

Scherer is probably on the right track, however, in focusing on the decisions by companies to adjust their R&D spending based on the expected payoff from it. The intensity of R&D spending is an increasing function of the rate of return a company expects to receive from performing it.

$$(2-11) \qquad\qquad S/Q = G(\rho^{exp}).$$

Companies in practice do not know what the future rate of return will be, but one indicator they use is the past behavior of productivity growth. As an approximation, therefore, we can assume that R&D intensity is a linear function (parameter ϕ) of the rate of multifactor productivity growth and other factors.

Review, vol. 73 (May 1983, *Papers and Proceedings, 1982*), pp. 215–18; and Frank R. Lichtenberg and Donald Siegel, "Using Linked Census R&D-LED Data to Analyze the Effect of R&D Investment on Total Factor Productivity Growth," Columbia University, January 1987.

63. Scherer, "R&D and Declining Productivity Growth," pp. 217–18.

(2-12) $S/Q = \phi\ (\Delta MFP/MFP) +$ other factors.

Given the uncertainty involved in technology development, companies typically set a fairly stable value for S/Q. The parameter ϕ is small. Big changes in the rate of productivity growth induce small responses in R&D spending decisions. When equation 2-12 is substituted into equation 2-10, therefore, there is a tendency for the reciprocal of ϕ to show up in the estimates of the coefficient of the rate of return (ρ). It will look as if small variations in S/Q have induced large responses in productivity growth.

The general point here is the one Scherer's analysis highlights. The econometric estimates are based on incomplete modeling. They assume that a correlation between R&D spending and multifactor productivity growth shows a causative link from R&D to growth. But in practice the correlation may be showing that the decline of technological opportunities led both to slower multifactor productivity growth and (somewhat) to smaller R&D spending.

Another point may be important in understanding the econometric literature, one linked to our earlier discussions of the electronics revolution and private versus social returns to R&D. The companies performing R&D in electronics may have successfully generated high *private returns* by developing and selling innovative products, even though the *social returns* to the innovations were low because the new equipment failed to pay off for the industries that invested in it. And electronics may not be the only example of this. As we see later, companies developing new technologies in electricity generation may have been quite successful in producing new equipment, even though the generating companies that installed it were unsuccessful.

If it is the case that the private return to R&D stayed high while the social return fell, then the econometric results do show that R&D during the 1970s generated productive scientific advances. There were *technological* opportunities. The problems came with the ways these advances contributed to overall growth.

Conclusions on the Contribution of R&D

Explanations of the slowdown conducted by Denison, Mohr, ourselves, and others have left a large puzzle that strongly suggests that slow innovation was a major reason for falling productivity growth. The analyses of innovation in the literature have looked at R&D spending but failed to come up with clear evidence that either the quantity of R&D spending, or the payoff to a given amount of R&D, fell by enough to explain much of this decline in growth. In fact the existing evidence points more toward the hypothesis that the payoff to R&D stayed high in the 1970s.

There are three possibilities. One is that the slowdown in productivity was caused by something that is unrelated to R&D or innovation. Another is that the technological opportunities did in fact decline, but that econometric or modeling problems obscured this. The third is that technology did well, but that the application of the technology to productivity was a failure. We now ask whether the experience of some specific industries can cast more light on these alternatives.

CHAPTER THREE

Industry Studies:
Chemicals and Textiles

TO LEARN MORE about the role of innovation in the productivity slowdown, we launched a research project to study the pattern of innovation and productivity in four major industries. This chapter reviews our results for chemicals and textiles,[1] and the next covers machine tools and electricity generation. This chapter also describes the approach we took to all four industries. Our hope was that by immersing ourselves in the specific experiences of each, we could find out what "really happened" to them. We wanted to avoid the problems that often plague studies based on econometric models, where precise quantitative estimates are obtained, but the reader is somehow left unconvinced.

Our efforts have left us with a more eclectic view. The complexity of the experiences of each industry and the diversity among the industries make it hard to generalize. Nevertheless, we have collected some specific quantitative evidence about innovation, evidence that we think is highly relevant to the productivity experiences of the industries, and from which we have drawn some lessons about the slowdown.

The four industries were chosen because of their mix of distinct characteristics. The chemical industry is high-tech, performing a large volume of research and development and relying on its own R&D for innovation. It is also capital-intensive. The textile industry, by contrast, is often described as low-tech because it performs little research and development, but it is perhaps better understood in relation to external technology sources. It relies on innovations made by its suppliers, principally the fiber producers (part of the chemical industry) and the textile machinery manufacturers. The textile industry is much less capital-intensive than is the

1. Most of the research and conclusions presented in this chapter appeared earlier in Martin Neil Baily and Alok K. Chakrabarti, "Innovation and Productivity in U.S. Industry," *Brookings Papers on Economic Activity, 2:1985*, pp. 609–32. (Hereafter *BPEA*.)

chemical industry, but it is about average relative to manufacturing as a whole.

The machine-tool industry performs some of its own research and technology development, but as a whole it does not spend a lot on R&D. While a few large companies develop sophisticated innovative machines, the many small companies focus mainly on custom-made machine tools for particular users, often in the automobile and aircraft industries. Because machine-tool companies often work with the auto companies in developing new generations of equipment, they rely on the R&D done in these purchasing industries. The machine-tool industry, like the textile industry, is in the middle range of capital intensity.

The electricity-generating industry is one of the most highly capital-intensive of any in the economy. Power plants require relatively few workers, although there is, of course, substantial employment in the industry associated with billing, distribution, maintenance, and administration. The industry does only a small amount of R&D itself, partly through cooperative research ventures such as the Electric Power Research Institute. The industry relies heavily on technology developed by equipment suppliers and by companies that design and build conventional and nuclear power plants.

Since the emphasis of our work on innovation is on explaining the slowdown in productivity, three of the four industries were chosen from the long list that experienced substantial productivity slowdowns in the 1970s; textiles was the one industry included that did not suffer a slowdown. We believed that if the pattern of innovation for this industry was found to be different from the others (and it was), this finding would add weight to the argument that innovation is important to productivity.

Innovation and Measured Productivity Change

When an invention emerges out of an initial search process or as a new concept, preliminary research must be done to develop it.[2] Not all inventions are developed, and for those that are, the bulk of the total research and development time and cost is incurred after the invention stage.[3] There

2. Major inventions are rare. Most ideas, inventions, and patents are applications of technologies that have been under development for some time or are already well established.
3. According to officials at the Du Pont company, 90 to 95 percent of R&D costs are incurred after the idea or invention has been formulated.

is a lag before a new product or process is ready for commercial introduction. It is at the point of commercial introduction that the new product or process is described as an innovation. After commercial introduction, the innovation diffuses through an industry, reaching, eventually, a point of maximum adoption.[4] Figure 3-1 summarizes the time pattern of the innovation process. The three lags are shown in the figure. During the periods of invention and development there is no productivity growth. In fact there may sometimes be negative effects on productivity as resources are put into the development process and prototypes are tried out with initially poor results.

Figure 3-1 shows three possible patterns of productivity growth during the diffusion period. Curve C shows a pattern often described in diffusion studies, in which the innovation is adopted slowly at first, then takes off, and then gradually reaches saturation. The underlying model used to describe such a pattern is based on the slow spread of information. Curve A shows a pattern that would be predicted by an economic-incentive model: the innovation is introduced first in uses where its increment to productivity is greatest. Later introductions face diminishing returns. Curve B is simply a middle ground.

In all three cases, A, B, and C, the innovation affects productivity *growth* when it is introduced. The *level* of productivity certainly does not jump at introduction, but productivity does grow as the new technology replaces the old. In case C, the effect on growth is small at first and then rises as the pace of diffusion speeds up. In case A, the effect on growth is greatest soon after introduction.

Lag patterns vary tremendously, depending on the nature of the innovation. When a major invention is made, there is usually a long period of development and a long period of diffusion. But the example of such major shifts is not necessarily applicable to econometric studies of the link from R&D to productivity, nor is it necessarily applicable to the innovations we will be studying. Major inventions are rare and often fail to have a big immediate effect on productivity. Instead, most of the R&D and the pro-

---

4. The relationships among R&D, patents, innovation, and productivity and the lags involved have been analyzed in a series of studies by Edwin Mansfield. A bibliography is given in Edwin Mansfield, "Microeconomics of Technological Innovation," in Ralph Landau and Nathan Rosenberg, eds., *The Positive Sum Strategy: Harnessing Technology for Economic Growth* (Washington, D.C.: National Academy Press, 1986), p. 324. See also Zvi Griliches, ed., *R&D, Patents, and Productivity* (University of Chicago Press, 1984); and Bronwyn H. Hall, Zvi Griliches, and Jerry A. Hausman, "Patents and R&D: Is There a Lag?" Working Paper 1454 (Cambridge, Mass.: National Bureau of Economic Research, September 1984).

Figure 3-1. *Time Pattern of the Innovation Process*

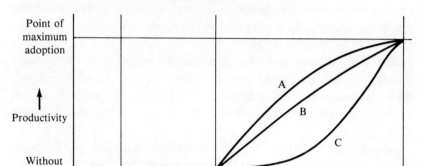

ductivity gains come over time as a whole series of new products and processes are developed that apply the invention in different areas.[5]

An example can illustrate this. The Du Pont company discovered nylon in the 1930s. The discovery was a major shift in technology, spawning literally thousands of new polymers, including plastics and synthetic fibers for clothing and industrial uses. Along with the new products there were innovations in process technology to make the new polymers and developments in machinery technology to use them. Since spinning and weaving productivity is higher with synthetics than with natural fibers, productivity in the U.S. textile industry in the 1970s was helped by the increasing proportion of synthetic fiber in its output.

Looking at nylon as a case study in innovation, one is impressed by the lags. There was a long development lag for nylon itself and then a very long lag before the full effects of the technological change were felt. Yet along the way there were many product and process innovations whose effects were much quicker. This flow of innovation was part of the process

5. A series of case studies of major innovations is described in L. Nabseth and G. F. Ray, eds., *The Diffusion of New Industrial Processes: An International Study* (Cambridge University Press, 1974).

of diffusion of the initial invention. The Du Pont staff told us that when a strategy for incremental innovation in process technology is devised, the lag from beginning the R&D to improved productivity in production can be as short as twelve months.[6]

The data on innovation that we collected consist of the incremental advances that are part of the spread of a major direction for advance, and such innovations are followed quickly by productivity growth. We are picking up innovation after the invention and development lags are over.

Innovations are divided into two broad categories: process and product innovations. Both are important to productivity growth, but because of the way productivity is measured, in practice they need to be distinguished.

Process Innovations

Process innovation is often thought to be the key mechanism for improving productivity. And in certain industries it is. Process-based industries, such as chemicals, develop new processes to enhance the productivity of their basic production. They also develop new processes to manufacture new products. However, new processes are also developed for other reasons, such as to save energy or lower levels of pollution, and these may not be reflected fully in measured productivity growth. In the case of pollution control, this is clear. The reductions in pollution levels are not priced and are not counted in productivity.

Reductions in energy use were not captured fully in productivity measures that used 1972 prices, since these prices included the pre-1973 price for energy.[7] The productivity data developed by the Bureau of Labor Statistics reported in chapter 1 should count such savings accurately. But some companies reported to us that the crisis atmosphere about energy in the 1970s had moved them to devote resources to energy-saving technology that were not justifiable on a cost-productivity basis. In that case, a shift to energy-saving process innovations from conventional productivity-enhancing process innovations could have reduced overall productivity growth.

6. The interviews conducted with Du Pont and other companies are described below.

7. A study by Resources for the Future estimates that the real price index for energy expenditures was 176 in 1977 and 263 in 1981, with 1972 equal to 100. These figures mean that each unit of fuel saved was being undercounted in productivity computations by a factor of 1.76 in 1977 and by a factor of 2.63 in 1981. See Resources for the Future, *A Historical Perspective on Changes in U.S. Energy-Output Ratios* (Palo Alto, Calif.: Electric Power Research Institute, 1985), p. 5-41.

Product Innovations

Process innovations are not the only or even the main source of productivity enhancement for U.S. manufacturing as a whole. New products supplied to an industry from outside and new products developed within an industry also improve productivity. The most obvious example is new equipment developed by the machinery industry and supplied to another industry. For example, in the 1970s new textile equipment developed by the European and Japanese machinery industries replaced the mechanical shuttle with air or water jets and brought about dramatic productivity benefits. Such innovations will always raise labor productivity. In practice they raise multifactor productivity also, because when the stock of capital of an industry is computed, the machinery is priced in such a way that part of the innovation in the machinery-producing industry is attributed to multifactor productivity growth in the machinery-using industry.[8]

New products manufactured within an industry do enhance its own productivity. In the period after a new product is introduced, productivity rises rapidly as the scale of production increases and the company moves down a learning curve. New products may also indicate new process developments. In the chemical industry, where product and process innovations are closely related, much of the R&D cost associated with a new product is spent on developing the process used to produce it.

As price indexes are now constructed by the Bureau of Labor Statistics, not all the productivity improvement resulting from new product development is reflected in official measures. Initially, a new product is excluded from the output price index. It is linked into the index once it is established. The dollar value of new products is counted right away, however, so that nominal output always includes new products. In computing real output for productivity purposes, therefore, nominal output, including new products, is deflated by a price index that excludes at least some of these products.

The product cycle is pretty much the same across most industries. Old products have a standard technology and their profit margins are gradually squeezed by competition. New products are typically introduced with high margins. These high margins are then effectively counted as high real output, so that the introduction of new products increases measured productivity. On the other hand, new products usually have declining relative

8. F. M. Scherer discusses the effect of innovation in supplying industries on productivity in using industries. See "Using Linked Patent and R&D Data to Measure Interindustry Technology Flows," in Griliches, ed., *R&D, Patents, and Productivity*, pp. 417–61.

prices, because of the learning curve and rapid productivity gains. And until the new product is linked into the price index, the *direct* effect of this price decline is missed. However, since a typical new product is just a variant of an old product, the old products do have to compete with the new products. The rapid productivity gains that occur for new products also hold down the prices of old products and, hence, reduce the increase in the industry price index even if it excludes these new products.

Without more information it is impossible to be sure of the size of the new-product bias in productivity measurement. New products do increase measured productivity, but the increase as now measured is understated relative to a true economic measure of productivity.[9]

Collecting the Innovation Data

Because the innovation process is so complex, it was hard to decide where to go to collect data on innovations and what innovations to include. Previous studies of innovation had emphasized only the big breakthroughs.[10] But small, incremental innovations can be equally important. A breakthrough innovation—the shuttle-less weaving machine—was available in the 1960s but could not be used throughout the textile industry until successive generations of new machinery had been developed that both perfected the technology and adapted it to produce the great variety of fabrics the industry makes. The new machinery developed over fifteen years or more made up a near-constant flow of equipment innovations that contributed to productivity growth.

Our decision was to collect a file of innovations for three of the four industries by surveying the technical trade periodicals that serve each.[11] New products developed within the industry and supplied to the industry are both advertised and described extensively in articles in these journals. New processes are also noted and described, if not always advertised. Although companies frequently keep secret the details of new process

---

9. Without implicating him in our conclusions, we would like to thank Jack E. Triplett, until recently at the Bureau of Labor Statistics, for a helpful conversation on this issue.

10. See, for example, Gellman Research Associates, *Indicators of International Trends in Technological Innovation* (Washington, D.C.: National Science Foundation, April 1976).

11. For chemicals, the periodicals covered were *Chemical Engineering, Chemical Engineering Progress, Chemical and Engineering News,* and *Chemical Week.* For textiles, the journals were *Textile World, Textile Industries, American Dyestuff Reporter, Textile Chemist and Colorist,* and *America's Textiles Reporter Bulletin.*

technology, the existence of a process innovation and some information about its character generally are reported. We double-checked our findings with industry engineers to learn whether our search procedure had missed significant innovations, particularly new process developments.

Our research assistants were graduate students at Drexel University who had at least undergraduate training in engineering. We trained them in the selection criteria described below and monitored their performance.

Although the assistants knew of the general goals of the project, no one involved had a preconceived view that innovation either had or had not slowed after 1973. In previous work, Baily had argued that factors other than innovation were responsible for the productivity slowdown.[12] Chakrabarti, who supervised the data collection, is primarily interested in the management aspects of innovation and how corporate strategies toward innovation are determined.

Following advice from chemical engineers at Drexel and in the industry, we established four categories of chemicals innovation (products, processes, equipment, and instruments) and set up criteria for judging whether a new item in fact represented something significantly new or improved. We tracked innovations originating both within and outside the industry. We judged new chemical products to be innovations if they were chemically new (that is, had new physical or structural properties), were significant modifications of existing chemicals, or were chemically reformed or recompounded for different applications. A new chemical process had to show changed inputs or yields or produce a new product. An equipment innovation, often incorporated into new processes, had to operate at new physico-chemical parameters or process new materials. A new instrument had to be able to measure the operation of chemical processes with greater precision, in a changed environment, or over a wider range.

The textile innovations were similarly classified and filtered, based on advice from textile engineers at the Philadelphia College of Textiles and Science. The industry develops a small number of process innovations for dyeing and finishing, which use process technologies similar to those in the chemical industry. The principal way in which innovations improve productivity in the textile industry is through new textile machinery. We judged new equipment to be an innovation if it showed improvements over existing equipment in such characteristics as speed of operation, ability to handle new materials, or reduced input requirements.

12. Martin Neil Baily, "Productivity and the Services of Capital and Labor," *BPEA, 1:1981,* pp. 1–50.

Instrument innovations in dyeing and finishing are also similar to those described in the chemical industry. Instruments are used, too, in spinning and weaving, where they can sense the characteristics of the fiber and control the machinery to allow more rapid operation and less breakage.

Innovations in textile material inputs, which consist of new fibers, finishes, and dyes, overlap with the chemical product innovations. Most of these inputs come from the chemical industry, although separate chemicals are frequently combined into finishes and dyes within the textile industry. Product innovations take the form of new yarns and fabrics.

Our coverage of innovations was rather comprehensive. For the years from 1967 to 1982, we found 574 process innovations and 2,773 new products in the chemical industry, 2,047 equipment innovations in textiles, and more than 8,000 innovations in the machine-tool industry. Once the innovations had been collected, we asked engineers to review the files, report on the completeness of our coverage, and rank the innovations by technical importance. We report, below, rankings on some of the more important innovation categories.

Innovation Patterns in Chemicals and Textiles

The chemical industry, defined as Standard Industrial Classification 28, but excluding the drug industry (SIC 283), achieved rapid multifactor productivity growth until 1973, when growth slowed substantially. Unadjusted for capacity utilization, it slowed even more after 1979, as shown in table 3-1.

If multifactor productivity is to be linked to innovation, however, it makes sense to adjust for capacity utilization changes. The rate of capacity utilization for the chemical industry has declined sufficiently over time, especially since 1979, to affect the measure of productivity. An adjusted growth rate for capital input to the industry can be constructed by multiplying the capital stock by the capacity utilization rate reported in Federal Reserve Board series for various industries. This procedure provides a better estimate of capital services actually used. The measure of multifactor productivity growth calculated from the adjusted capital input is shown in table 3-1. Even with the adjusted measure, it remains true that growth slowed substantially after 1973. But, as a result of the adjustment, one now sees that there was some recovery of growth from 1979 to 1983. Excess capacity in the chemical industry, it seems, brings about a substantial drop in productivity performance.

Table 3-1. *Multifactor Productivity Growth in the Chemical and Textile Industries, 1965–83*[a]
Percent per year

Industry and series	1965–73	1973–79	1979–83
Chemicals			
Unadjusted	3.09	1.73	0.98
Adjusted[b]	3.10	1.91	2.53
Textiles			
Unadjusted	2.61	3.37	3.18
Adjusted[b]	2.73	3.56	3.38

Source: Data provided by the U.S. Department of Commerce, Bureau of Economic Analysis, and U.S. Department of Labor, Bureau of Labor Statistics.

a. Multifactor productivity growth is calculated as the rate of growth of GDP originating in each industry minus the weighted average of the growth rates of the capital and labor inputs.

b. The adjusted growth rates for chemicals and textiles are calculated by multiplying the capital input by the Federal Reserve Board's industry-specific measure of capacity utilization.

Table 3-1 shows that the textile industry experienced no productivity slowdown at all after 1973. There was, in fact, some acceleration. Moreover, the capacity adjustment makes only a minor difference to the productivity numbers in textiles. One reason is that the industry had no widespread excess capacity, as reported in the Federal Reserve Board series. Another is that, since the industry is not very capital-intensive, a given amount of excess capacity has only a small effect on multifactor productivity.

Chemical Innovations

Table 3-2 reports the basic data on innovations in the chemical industry. The pace of innovation slowed considerably from 1967–73 to 1974–79, with the number of product innovations falling to only a fraction of the previous level, and instrument innovations falling almost in half. This is the kind of dramatic decline in innovation that one would expect to see, given the decline in productivity growth that took place in the industry.

The one series that does not fit well with the overall picture of declining innovation and productivity growth is the number of chemical process innovations. We suspected, based on other information, that the numbers in table 3-2 might be understating the decline after 1973 in process innovations that significantly enhanced productivity. Certainly the numbers are called in question by a U.S. Department of Commerce report on the plastics and synthetic materials industry (a major part of the chemical industry) that concludes: "A major factor underlying the evolution of the industry's

Table 3-2. *Innovations in the Chemical Industry, 1967–82*
Average number per year

Period	Chemical products	Chemical processes	Equipment	Instruments
1967–73	332.0	39.0	107.9	29.6
1974–79	38.8	32.3	56.7	18.2
1980–82	64.7	34.7	104.7	54.0

Source: Authors' computations. See text description.

input structure between 1958 and the early seventies was the great wave of cost-saving technical advances which swept over the industry before 1970. . . . However, it is also important to note that after 1977, the industry developed and widely dispersed only one major cost-saving innovation."[13]

We do not confirm the Commerce Department's statement exactly, but the idea that there was a wave of innovations in the 1960s that did not persist into the 1970s was confirmed for us by industry experts, and also by a recent major study of the synthetic fiber industry in the United States, Europe, and Japan, conducted by Anthony Cockerill.[14]

Did the results in table 3-2 understate the importance of the decline in process innovation? We examined the data more carefully to see whether the picture provided by the crude numerical count of innovations in the table was obscuring a significant change in the quality or the character of the innovations. Many of the process innovations in the file failed to enhance productivity in any clear-cut way. Many had an environmental aim—to reduce the toxic pollution emitted by existing process technologies. Many more changed the character or quality of the product. A number conserved energy. These last improved productivity, of course, but did not have a full impact on measured multifactor productivity.[15]

Table 3-3 gives data on the number of process innovations that can be identified as being productivity-enhancing, environment-related, or energy-related. It is striking both how few productivity-enhancing process

13. U.S. Department of Commerce, *The U.S. Plastics and Synthetic Materials Industry since 1958* (Washington, D.C., 1985), pp. 88–89.

14. Anthony Cockerill, "The Man-Made Fibres Industry: International Comparisons of Structure, Conduct and Performance," 2 vols., University of Manchester, Department of Management Sciences, September 1985, pp. 449–50.

15. There are three reasons for this. First, the productivity figures in table 3-1 are based on 1972 price weights. Second, the energy crisis led to a rush to save energy that was not necessarily cost-efficient even in current prices (the shadow value of energy conservation was high). Third, finding ways to save energy may have represented a long-term investment, not a short-run productivity payoff.

Table 3-3. *Process Innovations in the Chemical Industry, by Type, 1967–82*[a]

Average number per year

	Type of Innovation		
Period	Productivity-enhancing	Environment-related	Energy-related
1967–73	7.3	8.9	0.3
1974–79	4.2	7.0	3.5
1980–82	7.3	8.0	4.7

a. The innovations were classified by the authors. The three categories above are not exhaustive. The remaining process innovations were primarily to produce a new or modified product. A few innovations could not be classified.

innovations there were and how much greater their post-1973 drop-off was than the total decline in process innovations.

Although the table shows the importance of the environmental movement and antipollution efforts, it fails to provide evidence that they were a chief cause of the post-1973 slowdown. Over the whole period, more innovations were directed toward reducing pollution than toward productivity enhancement. But there were actually more environment-related process innovations before 1973 than there were from 1974 to 1979.

The energy-related innovations show exactly the pattern to be expected. There were almost no innovations in this area before 1973, but several afterward. The diversion of R&D effort toward saving energy may have curtailed innovations that save capital and labor.[16]

Our next step was to ask the chemical engineers who had advised us initially to go over the file to rank the innovations by technical importance, according to whether they were radical, or major; a significant improvement; or minor, or imitative.[17] As table 3-4 shows, the quality rankings do not change the picture much.[18] The data in the table on productivity-enhancing innovations slightly weaken the case that innovation declined, for they show that the falloff in minor productivity-enhancing innovations was greater than the decline in significant improvements. The falloff in radical and major innovations, however, was even larger than the decline

16. This hypothesis has been emphasized by Dale W. Jorgenson, "Energy Prices and Productivity Growth," *Scandinavian Journal of Economics*, vol. 83, no. 2 (1981), pp. 165–79.

17. Rankings were provided for us by Dipak Roy and William Herring, Amoco Chemical; Edward Hogan, PQ Corporation; Deepak Agarwal, Stearns Catalytic; and R. Mutherasan and Elihu Grossman, Drexel University.

18. The work of Samuel Hollander on the chemical industry indicates that small process innovations may be as important to productivity growth as large ones. See *The Sources of Increased Efficiency: A Study of Du Pont Rayon Plants* (MIT Press, 1965).

Table 3-4. *Process Innovations in the Chemical Industry, by Technical Importance, 1967–82*

Average number per year

	Productivity-enhancing			All processes		
Period	Radical or major importance	Significant improvement	Minor importance	Radical or major importance	Significant improvement	Minor importance
1967–73	0.6	4.1	2.6	3.1	22.7	13.1
1974–79	0.3	2.8	1.0	2.7	15.8	13.8
1980–82	0.7	5.7	1.0	2.3	23.7	8.7

Source: Authors' computations.

in total productivity-enhancing innovation shown in table 3-3. And when all process innovations are considered, the quality ranking strengthens the case for a decline. Minor innovations actually increased each year from 1974 to 1979.

We also obtained quality rankings for the chemical product innovations to see whether the overall decline in product innovation shown in table 3-2 might simply reflect a falling off of minor changes. Table 3-5 indicates that this was not the case. As one would expect, the vast majority of all product innovations were fairly minor, but all three rankings declined precipitously. Only one major product innovation occurred after 1973, according both to our file and to the engineer who did most of the rankings.

Although the period 1980–82 consists of only three years and encompasses two sharp recessions, the data nevertheless show signs of a recovery in process and equipment innovation in the chemical industry. Moreover, the productivity data in table 3-1 also seem to indicate some recovery of growth in this industry after 1979.[19] Thus the correlation between innovation and productivity performance evident from 1967 to 1973 and from 1974 to 1979 is continued after 1979. We are unwilling, however, to put much weight on such a short, turbulent period.

Textile Innovations

Table 3-6 reports the basic data for the textile industry, defined as SIC 22. The most important component of innovation for textile productivity,

19. See also table 1-3. It shows a partial recovery of productivity growth in chemicals from 1979 to 1985.

Table 3-5. *Product Innovations in the Chemical Industry, by Technical Importance, 1967–82*
Average number per year

Period	Radical or major importance	Significant improvement	Minor importance
1967–73	2.4	96.9	232.6
1974–79	0.2	9.0	29.7
1980–82	0.0	5.7	59.0

Source: Authors' computations.

new equipment, maintained continued vigor after 1973, the result of new generations of machinery that have consistently raised weaving and spinning speeds and reduced the labor required for restart.

The table does show some decline in the total number of process innovations. In the textile industry, however, in contrast to the chemical industry, when the process innovations were classified by type, as in table 3-3, we found no decline in the flow of productivity-enhancing process innovations. There were an average of 4.4 productivity-enhancing innovations a year during 1967–73, 5.0 a year during 1974–79, and 4.3 a year during 1980–82. The data on textile innovations, therefore, provide important support for the link between innovation and productivity performance. After 1973 the textile industry showed no slowdown either in productivity growth or in its two main production-related categories of innovation— equipment and productivity-enhancing processes.

The other innovation categories in table 3-6 do decline somewhat after 1973. The drop-off in instrument innovations is of only minor significance. Based on historical experience, the decline in the flow of new fibers could have been expected to be more important because the introduction of new fibers has figured prominently in textile productivity in the past. But the shift from natural to man-made fibers that has been crucial to productivity growth continued rapidly throughout the 1970s and thus diminished the significance of the decline in the number of new man-made fibers after 1973. This development is discussed in more detail below.

Two Case Studies: The Chemical and Textile Industries

At least for these two industries, the innovation data tell a clear story. To check this story further, we reviewed a variety of economic, business, and technical literature about the industries and also interviewed experts

Table 3-6. *Innovations in the Textile Industry, 1967–82*
Average number per year

Period	Equipment	Processes	Instruments	Fibers	Dyes and finishes
1967–73	134.5	17.4	53.9	19.0	267.0
1974–79	140.5	14.8	44.3	11.0	299.5
1980–82	154.3	9.3	37.7	4.0	180.0

Source: Authors' computations. See text description.

within the industries. The chemical industry was easier to study because of its excellent secondary sources and company staffs that were willing to talk about their own firms' histories. We interviewed a senior corporate planner, an R&D director, and a plant manager in each of three major companies—Du Pont, the Monsanto Company, and the Minnesota Mining and Manufacturing Company, 3M.

The Chemical Industry

Most output in the chemical industry consists of bulk or commodity chemicals.[20] Bulk chemicals include building-block chemicals, such as benzene and ethylene, that are used in later chemical processes; organic intermediates, such as formaldehyde and methanol; and final-use chemicals, such as fibers (nylon, polyester), plastics (polyethylene, polyvinyl chloride), inorganic chemicals (soda, chlorine), and fertilizers (ammonia, phosphates), that are shipped out of the chemical industry. The remainder of the output comes from specialty companies. We interviewed employees of two large producers of bulk chemicals, Du Pont and Monsanto, and of one specialty company, 3M.

BULK CHEMICAL COMPANIES. Employees at both Du Pont and Monsanto confirmed that productivity had slowed after 1973 and that the pace of innovation had declined in the 1970s. Both companies used internal productivity measures based on gross output per employee. Estimates of the slowing in the pace of innovation were more subjective. Virtually everyone with whom we spoke confirmed the existence of a slowdown, but R&D staff generally described it as less significant than did the pro-

20. Major secondary sources of information for this industry include First Boston Corporation, *Analysis of Chemical Production Capacities* (Boston, 1977); and Department of Commerce, *U.S. Plastics and Synthetic Materials Industry*. Innovation in the chemical industry has been studied extensively in the research program headed by Edwin Mansfield at the University of Pennsylvania.

duction staff. R&D staff blamed the slowdown on shrinking R&D budgets in the early 1970s. Other staff indicated that the R&D budgets were cut because the technological opportunities were limited. At the end of World War II immense technological opportunities awaited exploitation, partly as a result of developments stimulated by the war itself and partly because of a natural cycle in the technology of the industry, but by the 1970s research efforts were running into diminishing returns. A new impetus to technological development was needed.

As previously noted, Du Pont discovered nylon in the 1930s. Its commercial development was bound to take place over many years, and this development process was delayed by the war. The discovery spawned a whole series of innovations falling in two waves, one in the 1950s and one in the 1960s. These innovations were not limited to new fibers, because the new chemistry stimulated by nylon led to many other applications. Five thousand new polymers were developed in the first wave alone. In the 1950s and 1960s, process innovations went along with the new product development, and successive generations of new process technology resulted in large productivity gains. By the 1970s the waves of innovation had run their course, and the opportunities for rapid advance in existing product lines were limited.

Monsanto achieved rapid productivity gains until 1970 primarily by building larger and larger plants.[21] By the 1970s, however, the potential for scale-related design and materials innovations had largely been exhausted.

According to staff at both companies, innovation is closely linked to productivity growth in the long run. In the short run, organizational and managerial changes and improvements in work practices can make a substantial difference. But even these improvements come as part of the learning curve associated with new products and processes. If a company were to fail to make innovations, productivity growth would slow and stop after a few years. The consensus at both companies was that the slowing in the pace of innovation had contributed significantly to the slowdown in productivity growth. The interview responses thus provide independent support for the quantitative results given earlier.

When asked about other causes of the productivity growth slowdown after 1973, interviewees cited slow growth in product demand as being at least as important as the slowing of innovation. In neither company was

---

21. This interaction between innovation and economies of scale in the chemical industry has been analyzed by Richard C. Levin, "Technical Change and Optimal Scale: Some Evidence and Implications," *Southern Economic Journal,* vol. 44 (October 1977), pp. 208–21.

the slow growth in demand due primarily to the business cycle, although obviously the recession of 1974–75 did play a role. The two main causes cited for slow demand growth were structural—foreign trade and energy prices. The reduced number of product innovations itself also curbed demand growth.

Although the U.S. chemical industry does face direct foreign competition, the source of its trade difficulties was not its own foreign competition, but that of its customers. The textile and apparel industries, heavily affected by foreign competition in the 1970s, sharply reduced the growth rate of their purchases of chemicals, particularly synthetic fibers.[22] At the same time, petrochemicals suffered from the increased cost of oil and natural gas feedstocks, which raised final prices and caused a sharp reduction in demand growth.[23]

In both companies the view was that the change in the trend of demand growth had an impact on their productivity for as long as ten years. Interviewees acknowledged that they had been slow to realize that the trend had changed. Because of the several-year-long planning and construction period for new chemical plants, large-scale state-of-the-art plants, designed to achieve economies of scale and maintain or increase market share, continued to be brought on line even when company officials recognized that demand had fallen. These plants were then operated below capacity and thus inefficiently, production worker requirements in such plants being almost independent of output levels.[24]

SPECIALTY CHEMICAL COMPANIES. In addition to the producers of bulk chemicals, the chemical industry features smaller specialty companies. The large diversified companies, too, have specialty chemical divisions. We interviewed employees of 3M, a company that produces a great many different chemicals, about half of which are sold internally to its other divisions. Those interviewees described 3M's experience in the 1970s as being quite different from that of the large-scale producers.

The chemical operations at 3M are not very capital-intensive. A specialist in versatility, 3M is the sole source of many of its products, and in some cases it produces the entire year's output in one or two days. The

22. It should be stated for the record that Du Pont staff said that they were ready and willing to operate under a regime of free trade, provided they were free to sell fiber to whoever is producing textiles. Because of the Multi-fiber Arrangement and various rate restrictions, they are not allowed to compete freely overseas.

23. The recession cut demand after 1973, and then feedstock prices increased as price controls were lifted.

24. There is actually little direct labor used in a large chemical plant. The problem of excess labor apparently included additional sales and clerical staff hired in anticipation of sales volume growth.

equipment is then cleaned and used for another chemical. Because of the diversity of products, 3M's internal productivity numbers are of limited value, but employees judged that there had been little or no productivity slowdown after 1973. Pursuing their traditional innovation strategy of making continual product improvements or developing new products related to existing ones, they sensed no slowdown in the pace of those innovations in the 1970s. Nor did the company experience excess capacity after 1973. As demand growth slowed, 3M followed a strategy of minimizing investment expenditures and making more efficient use of their plants.

Although the experience of this specialty chemical company and that of the bulk producers differ sharply, they are consistent. The specialty company reported no slowdown in innovation, no slowdown in productivity, and no excess capacity.

OTHER CAUSES OF THE PRODUCTIVITY SLOWDOWN. Employees of the three companies also responded to questions about other possible causes of the decline in productivity growth. None thought that labor quality, work effort, or related labor issues had played an important role in changes in the trend of productivity growth. One company reported labor conflict occurring for a short period around 1975.

The problems associated with excess capacity are, of course, linked to capital services. But beyond this, none of the interviewees reported that accelerated obsolescence had been a serious difficulty for their company, although employees of all three cited examples tied to either regulation or energy costs.

The diversion of R&D resources to meeting environmental regulatory requirements or to saving energy was important for Monsanto and 3M, less so for Du Pont. From 1973 to 1979 half of Monsanto's R&D was environment- or energy-related. This draining of resources contributed to the decline in the number of cost-reducing innovations.

The Textile Industry

The U.S. textile industry is structured very differently from the chemical industry.[25] It consists of a large number of firms (6,000 in 1973),[26]

25. Major secondary sources for this industry include Brian Toyne and others, *The U.S. Textile Mill Products Industry: Strategies for the 1980's and Beyond* (University of South Carolina Press, 1983); and National Research Council, *Competitive Status of the U.S. Fibers, Textiles, and Apparel Complex* (Washington, D.C.: National Academy Press, 1983).

26. Jordan P. Yale, "The Textile Industry in Transition," Report 532 (Menlo Park, Calif.: Stanford Research Institute, November 1974), p. 1.

most of them small. The largest, Burlington Industries, had only 5 percent of the market in 1979 and the next largest, J.P. Stevens, had 3 percent. Plant sizes, too, are often small: in 1979, 70 percent of the plants had fewer than 100 employees.[27] Many of these small plants and companies subcontract for the larger companies in a process that is called "jobbing" —performing such specific tasks as finishing the woven fabric.

We interviewed R&D directors and plant managers of two of the larger companies in the industry, as well as staff at the Philadelphia College of Textiles and Science and the Textile Research Institute in Princeton. The latter were able to provide information about both large and small companies in the industry. Partly because many textile companies are privately owned and release little information, and partly because of political negotiations that were proceeding on textile quotas at the time of this study, the companies asked not to have their names associated with particular views.

Both industry personnel and secondary sources agree that the rate of innovation and productivity growth in textiles remained strong from 1973 to 1979. Again, there is confirmation of the pattern we reported in our own data.

The industry experienced rapid output and demand growth until the late 1960s and early 1970s, when foreign competition began to erode market growth. Textile exports were reduced, and imports of foreign-made apparel increased.[28] The textile industry itself, however, still had a trade surplus in 1981.

Unlike the chemical industry, the textile industry adapted to the change in growth without creating persistent excess capacity. The older and less-efficient plants were closed, and employment fell rapidly. One company reported that it had actually increased its capital utilization in the 1970s, moving to seven-day-a-week, three-shift operation to minimize investment requirements. Differences in the technology account for the ability of the textile industry to adjust to lower output growth.[29] The companies do not have large capital-intensive process-based plants and do not build much ahead of demand. Textiles and chemicals also have different traditions of labor relations; textile companies do not engage in much labor hoarding.

27. Toyne and others, *U.S. Textile Mill Products Industry,* pp. 5-1a, 5-22.
28. There has not been free trade in textiles and apparel in the United States. Restrictions have been placed on U.S. imports that limited the erosion of the market for U.S. textile manufacturers.
29. The textile industry has high average variable costs and low fixed costs relative to the chemical industry.

One of the important sources of rapid and continuing productivity growth cited by interviewees—the shift from natural to man-made fibers—was not well reflected in the innovation data. The production of textiles from natural fibers requires many steps before the yarn is woven. For cotton fabrics, for example, the cotton bales must first be broken up and the cotton fibers loosened and then blended. In the next step, "cording," the fibers are cleaned and formed into strands. Three more steps take place before spinning: "drawing" the fibers, "winding" the strands, and "roving," which reduces the size of the strands and winds them into fibers ready for spinning. After spinning, the yarn must still move to the "winding and warping" step before it is ready for weaving.

Synthetic fibers, provided by the chemical industry, often bypass the stages before spinning and are sometimes even spun in chemical plants before being sold. Moreover, because of their special physical properties, synthetic fibers can be spun and woven at high speeds. These efficiently produced new fibers took over the market. In 1957 cotton and wool made up 71 percent of all U.S. fiber consumption. The proportion had fallen to 56 percent by 1966 and to 24 percent by 1981.[30] Textile industry productivity growth was thus boosted throughout the 1970s by the waves of product innovation in the chemical industry that took place in the 1950s and 1960s. It is worth noting that foreign trade contributed to this productivity enhancement. In 1981 the United States had a substantial net trade surplus in man-made yarn and fabric and a substantial deficit in cotton fabric. The U.S. textile and apparel industries have maintained their comparative advantage most effectively in synthetic fabric.

Successive generations of new equipment have also continued to raise textile productivity growth. Carrying out only a small amount of R&D itself, the industry has had access to a constant flow of new technologies, generated primarily from equipment suppliers. The latest Swiss Sulzer automatic looms can now weave 760 meters of denim a minute, nearly four times the speeds achieved twenty years ago.[31] Developments in spinning technology have been just as dramatic. Air jets are replacing spindles, just as they earlier replaced shuttles. Circulating air currents now spin the yarn into thread.

Other developments in the industry that might have affected productivity were reported by the interviewees. Old capital became rapidly obsolete

30. U.S. Bureau of Labor Statistics, *Technology and Manpower in the Textile Industry of the 1970's,* Bulletin 1578 (Washington, D.C.: Department of Labor, August 1968), p. 40; and Toyne and others, *U.S. Textile Mill Products Industry,* p. 5-7.

31. Figures are from "Denim: Jeans Fit for Robots," *Economist,* July 27, 1985, p. 82.

in the 1970s as new equipment was installed and old plants closed. Labor quality and work effort had no major influence on productivity trends. Increased foreign competition encouraged management to introduce more productive work practices, benefiting productivity in the past ten years. The industry has been affected by economic regulation, particularly by Occupational Safety and Health Administration requirements for a clean work environment. Regulation did not cause a particular productivity slowdown in the 1970s, however, because the regulations in this area were introduced in the 1960s and then strengthened over time. Finally, the new generations of machinery are more energy-intensive than their predecessors.

Conclusions

Experience in the chemical and textile industries strongly supports the idea that productivity performance and innovation are linked. As for the other possible suspects discussed in chapter 2, none of our interviewees argued that a collapse of work or skills had been a major factor. Nor did anyone argue that capital obsolescence caused the slowdown; in fact the evidence went the other way, because the textile industry did discard capital but nevertheless had excellent productivity growth. Looking at the effect of demand, we found both industries suffered weak demand and one did well and the other badly, so that on the face of it weak demand does not look like a contributor to the slowdown. However, there is a key difference evident in the responses of the two industries to demand. Because the chemical industry is process-based and uses large fixed plants, it had trouble adjusting to slow growth. By contrast, textile firms shut down their inefficient plants and fired excess workers, so weak demand actually concentrated production in the best plants.

CHAPTER FOUR

Industry Studies:
Machine Tools and Electricity

THE TWO industries we chose for the second round of our case studies—machine tools and electric power—have without question experienced severe declines in productivity growth in recent decades. Basically, productivity growth in machine tools did well until 1977 and then deteriorated; in electric power, most analysts agree that growth began to decline even earlier, in the 1960s. Details of the slowdown in productivity growth for the two industries are shown in tables 4-1 and 4-2. The data for utilities include gas and water, but electric power generation is the largest segment of the total.

Again we asked whether the rate of productivity growth may have been pulled down by a slowing in the pace of innovation. Our approach to analyzing innovation in the machine tool industry was essentially the same as that used for chemicals and textiles. There is a single comprehensive periodical, *Tool and Production,* that reports on innovations in the industry, and data on more than 8,000 innovations were collected. Though some of these were later excluded as trivial, the base of innovation data remained fairly comprehensive. Our approach to electric power was different. For this industry, we used secondary sources rather than collecting our own primary data to evaluate the way in which technology has evolved in the industry. We did not develop a year-by-year tabulation of innovations. This change in strategy was brought about partly as a result of constraints on our own resources. The collection and processing of the innovation data proved to be extremely time-consuming. But even with unlimited resources, we might still have altered our approach because this industry is one in which the success or failure of two key technologies was vitally important, namely, the development of supercritical turbines and the shift to nuclear power.

Table 4-1. *Productivity Growth in Public Utilities and Electricity Generation, Selected Periods, 1953–83*

Percent per year

Industry and study	Period	Percent	Period	Percent
Public utilities				
		Labor productivity growth		
Gordon, BEA	1953–58	5.27	1968–73	2.60
	1958–63	6.17	1973–78	1.10
	1963–68	4.81	1978–81	−1.98
		Multifactor productivity growth		
Kendrick	1953–57	5.60	1966–69	2.80
	1957–60	5.30	1969–73	0.00
	1960–66	3.50	1973–79	−0.90
Electricity generation				
		Labor productivity growth		
BLS	1958–63	7.70	1973–78	1.82
	1963–68	6.36	1978–83	−1.28
	1968–73	4.64		
		Multifactor productivity growth		
Nelson and Wohar	1956–61	2.69	1968–73	2.19
	1962–67	2.74	1974–78	−0.42
Gollop and Jorgenson	1957–60	3.96	1966–73	−0.35
	1960–66	2.07		
Gollop and Roberts	1973–74	−3.27	1976–77	−0.72
	1974–75	−2.95	1977–78	0.17
	1975–76	−2.71	1978–79	1.39

Sources: For public utilities—Gordon-BEA: Robert J. Gordon, "The Productivity Slowdown in the Steam-Electric Generating Industry," Northwestern University, February 1983. Gordon uses data from the Department of Commerce, Bureau of Economic Analysis. Kendrick: John W. Kendrick, *Interindustry Differences in Productivity Growth* (Washington, D.C.: American Enterprise Institute, 1983), pp. 6–7.

For electricity generation—BLS: Bureau of Labor Statistics; Nelson and Wohar: Randy A. Nelson and Mark E. Wohar, "Regulation, Scale Economies, and Productivity in Steam-Electric Generation," *International Economic Review* (February 1983), p. 59; Gollop and Jorgenson: Frank M. Gollop and Dale W. Jorgenson, "U.S. Productivity Growth by Industry, 1947–73," in John W. Kendrick and Beatrice N. Vaccara, eds., *New Developments in Productivity Measurement and Analysis* (University of Chicago Press, 1980), p. 119; Gollop and Roberts: Frank M. Gollop and Mark J. Roberts, "Environmental Regulations and Productivity Growth: The Case of Fossil-Fueled Electric Power Generation," *Journal of Political Economy*, vol. 91 (August 1983), p. 670.

Machine Tool Innovations

The most important category of machine tools includes those that cut metal.[1] The basic machines of this type are milling, drilling, and grinding machines and lathes. The principal recent development in the industry has been the adoption of more advanced technology in machines that perform these cutting functions. These are machining centers that use automatic

1. In recent years such machines have also been designed to cut plastics and other materials.

Table 4-2. *Multifactor Productivity Growth in the Machine Tool Industry, 1965–82*[a]
Percent per year

Item	1965–73	1973–79	1979–82
Unadjusted	1.15	−0.30	−7.60
Adjusted[b]	2.39	−0.53	−2.26

Source: Calculated by the authors using data from the U.S. Department of Commerce and the U.S. Department of Labor, Bureau of Labor Statistics.

a. Multifactor productivity growth is calculated as the rate of growth of GDP originating in each industry minus the weighted average of the growth rates of the capital and labor inputs.

b. Calculated by using a regression in which the change in hours is used as a cyclical variable, since no capacity utilization measure is available.

tool changers, machines that use lasers to cut, and most important, numerically controlled and computer numerically controlled machines. While the new machines perform the same kind of cutting functions as the basic equipment, they rely less on the speed and skill of the operator. They can work more quickly and do more complex tasks.

The second category of machines includes those that shape or handle rather than cut metal. These are forming, handling, welding, and stamping machines. The main demand for these machines comes from auto and appliance companies that use sheet metal in production.

Like the equipment innovations in textiles, the new machine tools reported were judged to be innovations if they showed improved speed of operation, reduced materials requirements, or reduced input requirements.

Table 4-3 shows the annual flow of cutting-machine innovations for the eight subcategories. Within these cutting-machine innovations a fairly marked pattern emerges. The pace of innovation fell off after 1970 and then picked up after 1977. These periods were used rather than the pre-1973 and post-1973 periods shown for chemicals and textiles because the turning points in the flows of innovations occurred at times that did not match up exactly with the turning points in the productivity data. The pattern is different from the one we found in chemicals.

Taken individually, numerically and computer-controlled machinery innovations have patterns that are different from the rest. Together, though, the combined flow of innovations of the two types showed a marked lull from 1971 to 1977 in common with other cutting-machine innovations. (Computer control has taken over from numerical control alone.)

Table 4-4 shows the flow of innovations in forming, handling, welding, and stamping machines. There is some evidence of a 1970s decline in the

Table 4-3. *Innovations in Cutting Machines, by Type of Machine, 1967–83*

Average number per year

Type of machine	1967–70	1971–73	1974–77	1978–80	1981–83
Milling	15.8	7.0	6.5	15.7	14.0
Drilling	26.8	15.7	12.3	32.0	20.3
Grinding	5.3	3.0	1.8	36.7	40.7
Lathes	16.8	8.3	11.0	28.3	15.7
Machining centers	0.5	0.7	2.5	12.3	15.0
Lasers	3.8	3.3	0.3	5.0	9.7
Numerically controlled	16.3	8.3	9.0	4.0	0.3
Computer numerically controlled	...	...	2.0	17.7	19.7

Source: Authors' calculations based on data from *Tool and Production*.

flow of innovations in forming and handling machines, but no such evidence in welding and stamping machines.

The quantitative evidence in tables 4-3 and 4-4 reports the main effort of the data collection process for this industry. And cutting machines represent the bulk of product shipments for this industry. Innovations in other areas were also noted, however.

—A rapid flow of new controls was introduced in the late 1970s and early 1980s. Rather than buying numerically or computer-controlled machinery from scratch, many machine tool users converted other designs using controls.

—The number of innovations in robot and optical machine tools was small over the period but had begun to increase by the 1980s.

—Many innovations in cutting materials and components were introduced over the period but with no clear time pattern.

When our findings for the chemical and textile industries were first released, skeptics suspected that a little alchemy had been performed to help generate the striking correlation between the innovation series and the productivity series. Clearly we used no alchemy on the machine tool data. There was a lull in innovation in cutting machines in the early 1970s, but taken as a whole, tables 4-3 and 4-4 show that the correlation between innovation and productivity is not as close as in the two earlier cases. The time-series data on innovations do not confirm a direct link from slow innovation to slow productivity growth. Nevertheless, we will argue that the performance of this industry was in fact tied to its pattern of innovation.

Table 4-4. *Innovations in Shaping and Handling Machines, by Type of Machine, 1967–83*
Average number per year

Type of machine	1967–70	1971–73	1974–77	1978–80	1981–83
Forming	96.0	50.7	36.3	52.0	32.3
Handling	80.8	51.3	50.0	78.3	53.0
Welding	3.8	9.0	11.3	16.7	12.7
Stamping	17.0	19.7	15.5	24.0	20.7

Source: Authors' calculations based on data from *Tool and Production*.

The Machine Tool Industry

The U.S. machine tool industry consists of many small establishments and a few larger ones.[2] In 1972 there were 1,232 companies with 1,277 establishments in the industry. Total employment was 76,600, representing on average 62 workers for each firm and 60 for each establishment. Only two establishments had 2,500 or more employees. Only seventy-one establishments had 250 or more employees. The average size of firms and establishments was smaller in 1972 than it had been in 1967, and it had declined further by 1982. In 1963 there were on average 73 employees per firm. This figure had fallen to 61 by 1982.[3]

Citing average firm size exaggerates the importance of small firms, of course, because the large firms have a disproportionate share of employment. But even so, the picture of a small-company industry remains clear. In 1972, 79 percent of employment was in establishments with fewer than 1,000 employees. By 1977 only one company had 2,500 or more employees.[4] The industry produces specialized tools for particular applications and often uses labor-intensive production techniques.

Table 4-5 summarizes the main technology changes in the industry as

2. This section extensively draws upon material prepared by our assistant Nathaniel Levy. See "The Machine Tool Industry," Brookings, Spring 1986. Other important sources include U.S. Bureau of Labor Statistics, *Outlook for Numerical Control of Machine Tools*, Bulletin 1437 (Washington, D.C.: Department of Labor, March 1965), and *Technology and Labor in Four Industries*, Bulletin 2104 (Washington, D.C.: Deparment of Labor, January 1982); Horst Brand and Clyde Huffstutler, "Trends of Labor Productivity in Metal Stamping Industries," *Monthly Labor Review*, vol. 109 (May 1986), pp. 13–20; National Machine Tool Builders' Association, *Economic Handbook of the Machine Tool Industry* (McLean, Va., various years); *American Machinist*, various issues; and David J. Collis, "The Machine Tool Industry and Industrial Policy, 1955–82," in A. Michael Spence and Heather A. Hazard, eds., *International Competitiveness* (Ballinger, 1988), pp. 75–114.

3. NMTBA, *1983–1984 Economic Handbook*, p. 69.

4. Ibid.

Table 4-5. *Major Recent Technology Changes in Metalworking Machinery*

Technology	Description	Labor implications	Diffusion
Numerically controlled machine tool (NC)	Tool is controlled by instructions received from tape, punched cards, plugs, or other media. Allows rapid change to new product designs; permits stricter tolerances for parts; and reduces setup time. Useful for small batch production, because parts can be machined by merely changing tapes and resetting tool.	Estimated reduction in machining time of 35–50 percent; typically used two or three shifts. Reduces unit requirements for machine operators; requires less skill than manually operated tools, creates new job of programmer, requires more broadly trained maintenance personnel.	Three percent of all machine tools in metalworking are NC, but they account for a much larger proportion of total output. In metal-cutting sector, the value of NC machines is estimated at 30 percent of all machine tools installed in 1979. Considerable growth in the number of NC tools and their share of total output is expected in the 1980s.
Numerical control by computer (CNC)	On-board computer stores and conveys information directly to NC control unit; utilizes latest microprocessor technology.	Same labor implications as NC but, unlike NC, may require computer personnel. Saves time in reprogramming to remove errors or make design changes. Requires maintenance personnel with electronic skills.	Significant proportion of NC machines; mainly limited to larger and medium-size machine tool shops. Expected to increase substantially in the 1980s as result of cost reductions in electric controls; some smaller shops will introduce it.
Machining center	An automatic tool changer makes the center a multifunction NC machine. Each center is equivalent to several machines, each having a specific function.	Raises productivity by permitting operations on many surfaces of a part in a single setup. Operator may control several machines.	Accounts for a small but growing percentage of machine tools in larger plants, but a disproportionately large share of the industry's output.
Adaptive control	Automatically controls feed rate to reduce or eliminate such factors as vibration, tool wear, and cutting temperatures, and alerts operator. Can be used with conventional tools or with NC.	Raises productivity in machining through substitution of sensors for workers' own perceptions. Reduces skill requirements.	Used by large plants. Utilization by small shops will depend upon development of improved sensors and their cost effectiveness relative to the availability of skilled workers and also the type of work performed.

Technology	Description	Effect on labor	Diffusion/use
Computer-aided design/computer-aided manufacture (CAD/CAM)	Computers are used to develop designs for products to be manufactured (CAD). CAM directs numerically controlled machines and automatically guides workpieces among machines on computer-controlled handling systems.	Reduces need for low-skilled operators; increases requirements for higher skilled workers.	Used by large machine tool manufacturers only; diffusion to medium-size firms will be severely limited by the technology's cost.
Digital readout (DRO)	A device is applied to movable portion of a machine tool to measure its actual movement; can provide some automatic control; measurement appears on a display unit.	Operator efficiency and accuracy are enhanced during the positioning phase of the machine cycle. Operators are trained in less time and fatigue is reduced.	Use is limited but increasing among machine tool builders; already widespread among tool-and-die shops. Producers of DROs expect a 25 percent annual growth in their sales to the metalworking machinery industry in the next several years.
Manual-data-input control (MDI)	Enables an operator to change the position of a machine automatically; also identified as "operator-programmed NC."	Machinist can plan and enter part programs: possible in "shop language." Training period shorter than for NC programming.	Precise data on utilization are unavailable, but its use is spreading among machine tool builders and even more rapidly in the contract tool-and-die shops.
Cutting-tool materials	Durable new materials, such as coated carbides, polycrystalline diamonds, and special ceramics meet continued increases in machining speed.	Reduce labor requirements somewhat because tools do not have to be changed as often.	Tungsten carbide expected to remain the major material, but coated carbides may increase from current 15 percent to 25 percent of all cutting-tool materials in metalworking machinery in 1985.
Group technology (GT)	Management skills used to reduce small batch operations. Involves the grouping of parts on the basis of similar shapes or processing requirements. Workers may perform a wide range of tasks.	Improves efficiency and quality of output. Workers may broaden skills and replace narrow specializations.	Used by some large machine tool builders; elements of GT likely to spread slowly to smaller builders and tool-and-die shops.

Source: U.S. Bureau of Labor Statistics, "Technology and Labor in Four Industries," Bulletin 2104 (Washington, D.C.: Department of Labor, January 1982), p. 25.

Table 4-6a. *Shipments of Machine Tools by U.S. Companies, Selected Years, 1957–83*
Billions of 1983 dollars

Year	Metal cutting	Metal forming
1957	4.8	1.8
1960	2.7	1.1
1965	4.4	1.9
1968	5.0	1.5
1973	3.2	1.3
1979	4.1	1.3
1981	4.5	1.1
1982	3.0	0.7
1983	1.4	0.5

Source: National Machine Tool Builders' Association, *1984–1985 Economic Handbook of the Machine Tool Industry* (McLean, Va., 1983), p. 42.

described by the U.S. Bureau of Labor Statistics. Numerical control of machine tools, modified in the 1980s with the development of microcomputers, promised, in principle, tremendous productivity gains both to users outside the industry and to machine tool companies themselves, who use their own industry's products in production. In particular, the specialized products, often single-order products, that the industry manufactures are just what computer-controlled equipment can handle well. Design specifications can be programmed into the computer built into the machine tool.[5] As the compilers of the table note, however, the technology has been adopted very slowly. The surveys of machine tools by the journal *American Machinist* confirm this. In 1968, 0.5 percent of all the machine tools in use in metalworking industries were numerically controlled. This figure rose to 0.9 percent by 1973, to 2.0 percent by 1976–78, and to 4.7 percent by 1983.[6]

Demand for the machine tool industry's products is highly cyclical and has become very sensitive to imports. In addition, the industry uses highly skilled labor and is reluctant to release workers during downturns because the workers develop firm-specific skills and cannot easily be replaced. When demand is variable and employers are reluctant to cut employment, productivity becomes volatile. The weak productivity in the industry in the

5. For an extended discussion of the potential of this technology, see U.S. Office of Technology Assessment, *Computerized Manufacturing Automation: Employment, Education, and the Workplace* (Washington, D.C.: OTA, April 1984).

6. NMTBA, *1983–1984 Economic Handbook*, p. 261.

Table 4-6b. *U.S. Production, Imports, Exports, and Domestic Consumption of New Complete Machine Tools, Selected Years, 1965–83*

Millions of current dollars

Year	Production	Exports	Imports	Consumption	Price index
1965	1,430	222	56	1,264	22.8
1973	1,788	325	167	1,629	33.3
1977	2,453	427	401	2,428	54.8
1979	4,064	619	1,044	4,489	70.2
1980	4,812	734	1,260	5,338	82.0
1981	5,111	950	1,431	5,593	90.8
1982	3,749	575	1,218	4,392	97.4
1983	2,145	359	921	2,707	100.0

Source: NMTBA, *1984–1985 Economic Handbook*, pp. 41, 126.

1980s has been attributed to the tremendous declines in output.[7] The cyclical adjustment of the productivity data given in table 4-2 may not have captured the full impact of weak demand on productivity.

Table 4-6 shows output, export, and import data for the industry. Although the industry has always been vulnerable to cyclical downturns, the collapse of demand during the 1980s was absolutely unprecedented. Much of it was a consequence of the overvalued dollar. Imports had a direct effect on the industry as they took 34 percent of the domestic market in 1983, up from 23 percent in 1979. In addition, the metalworking industries that buy machine tools were hard-hit, so that total consumption fell 73 percent between 1981 and 1983 (82 percent in constant dollars).[8]

The overvalued dollar of the 1980s was tremendously important to the trends of imports and exports. But technology was just as important: imports rose rapidly from 1965 to 1979 when the dollar was *not* consistently overvalued. Exports could have been sustained more effectively if U.S. machine tools had retained the technological edge that they had in the 1960s. And the machine-tool-using industries would have kept their demand for all machine tools stronger if they had been convinced that the developments in computer-controlled machinery really offered great productivity gains.

The United States, Germany, and Britain were the dominant powers in

7. See Brand and Huffstutler, "Trends of Labor Productivity," for a discussion of these issues.

8. NMTBA, *1983–1984 Economic Handbook*, pp. 41, 126. Calculated using natural logs.

the industry after World War II, especially once Germany had reestablished its industry. The U.S. position was extremely strong; it was at the leading edge of technology. Britain tended to license or copy its technology from the U.S. companies. Over time, low-wage countries developed expertise in the production of basic machine tools, and they were able to export them competitively by the 1970s. Japan began as one of the successful low-wage competitors, but its relative wage had risen well above the average of the developing countries by the 1970s. This meant that during the seventies the United States, Japan, Germany, and Britain were all seeking ways to sustain their positions. Britain failed to do so and has had to contract its industry sharply. Germany has not made the transition to the new computer-controlled technology but has been able to move upscale within the existing technology, making high-quality tools that embody incremental advances over older machinery. Germany has also been able to take advantage of cooperative industry R&D programs to overcome the problems created by an atomistic industrial structure.

The Japanese have been extraordinarily successful in taking advantage of the opportunities for major technological advances in computer-controlled cutting machines, laser-cutting techniques, and robots. They have also been successful in moving upscale within the existing technology.

The U.S. industry had a strong technology base in the 1960s, and since then a few companies have successfully matched either the German or Japanese strategies. The industry as a whole, however, has failed to keep pace with the speed of worldwide technology development. This failure was particularly evident in the lull in innovation in the early 1970s, when foreign competitors were pushing ahead.

As noted earlier, cutting machines make up the central product group of the machine tool industry, and there was a reduction in the flow of U.S. innovations in this area from 1971 to 1977. It is likely that this weakness of innovation hampered productivity growth, but the decline was obscured for a time by the strong growth in product demand until 1973.

We found no evidence that the lull in innovation in the U.S. machine tool industry resulted from any exhaustion of technological opportunities. This conclusion is reinforced by the finding that the U.S. industry scrambled hard to innovate after 1977, realizing that European and Japanese machinery suppliers had developed more advanced products. Unhappily, the American effort to keep up or catch up appears to have failed, partly because the overvalued dollar greatly exacerbated the industry's competitive problems.

A rapid flow of innovations after 1977 was not accompanied by rapid productivity growth because demand was weak for the products developed by the U.S. industry. The profitability of the industry was also weak, impeding its ability to invest and equip itself with the most advanced machines.

Slow productivity growth was caused by a failure to capitalize on the advances in technology that were taking place in other countries. This failure led directly to slow growth and created serious structural weakness in demand as other countries took over the U.S. market.

Innovation and Productivity in Electricity Generation

Until the mid-1960s the electricity-generating industry was one of the stand-out performers in productivity growth. In fact the regulated monopolies—public utilities and telephone communications—both achieved growth rates far in excess of the national average. This favorable experience continued for the telephone industry, but growth collapsed in the 1970s in electric and other utilities.

There were four main causes of this dramatic decline in economic performance. First, excess capacity developed after 1968 and got worse after 1973. Second, environmental regulations led to fuel switching and to capital expenditures that failed to contribute to measured productivity. Third, utilities switched to coal and nuclear fuel in response to the jumps in oil prices. Fourth, the potential for productivity growth through increases in scale was largely exhausted by the 1960s, and the main potential sources of further productivity gains through innovation were unsuccessful.

Capacity

We constructed an index of capacity utilization for the industry, defined as the ratio of electricity generated (in kilowatt-hours) to installed capacity (in kilowatts). This utilization rate is shown in table 4-7, with a base year of 1968. After fluctuating around 100 from 1950 to 1969, the rate declines substantially and is down to 81 percent by 1980. A particularly sharp drop occurs in 1973–75, and this corresponds strikingly with the three disastrous years for multifactor productivity growth shown in the Gollop and Roberts data in table 4-1.

The drop in utilization between 1969 and 1973 was caused by rapid capacity expansion by utilities. The demand for electricity grew at around 7 percent a year from 1950 to 1968 and fell only trivially, to 6 percent a

Table 4-7. *Index of Capacity Utilization in Electricity Generation,*
Selected Years, 1950–80
1968 = 100

Year	Rate	Year	Rate
1950	101	1975	82
1958	97	1976	83
1963	95	1977	82
1968	100	1978	82
1973	92	1979	81
1974	86	1980	81

Source: Calculated by the authors as described in the text from data in Robert L. Loftness, *Energy Handbook*, 2d ed. (New York: Van Nostrand Reinhold, 1984).

year, from 1968 to 1973.[9] Some utilities experienced brownouts during peak demand periods, particularly as air conditioning spread rapidly. In short, they were under great pressure to expand capacity.[10] Yet just when they succeeded in doing this, in 1973, fuel prices jumped and demand growth fell to 3 percent a year over 1973–78, and fell again to 1½ percent a year over 1978–80.

Like the chemical industry, the electricity-generating industry found the adjustment to unanticipated slower demand growth extremely difficult. There was a lag before it was realized that the trend had shifted, and then a long lag before capacity could be adjusted downward. It takes several years to build a conventional plant and even longer to build a nuclear facility. The 20 percent reduction in capacity utilization between 1968 and 1980 represented a loss of capital services that adversely affected multifactor productivity growth. With capital representing 40 percent of income, the loss in the level of productivity was 8 percent by 1980. If excess capacity lowered efficiency, the loss was even greater.

Regulation

In 1970 the Clean Air Act Amendments were passed.[11] They established uniform national standards through the Environmental Protection

9. Robert L. Loftness, *Energy Handbook*, 2d ed. (New York: Van Nostrand Reinhold, 1984), p. 152. Calculated using natural logs.

10. For a discussion of this issue, see Andrew S. Carron and Paul W. MacAvoy, *The Decline of Service in the Regulated Industries* (Washington, D.C.: American Enterprise Institute, 1981).

11. This section draws on Frank M. Gollop and Mark J. Roberts, "Environmental Regulations and Productivity Growth: The Case of Fossil-Fueled Electric Power Generation," *Journal of Political Economy*, vol. 91 (August 1983), pp. 654–74.

Agency for six pollutants. One of the chief pollutants falling under the act was sulfur dioxide. And in 1970 most sulfur dioxide emissions were produced by electric power plants.

The restrictions embodied a form of grandfather clause, under which plants built before August 1971 are subject to state regulations. States vary widely in the extent of enforcement and the mechanism used. Some require the use of low-sulfur fuel, others concentrate on the emission level itself, limiting the number of pounds of sulfur dioxide per million BTUs. Some states apply more stringent regulations to larger plants.

Plants built or modified after August 1971 are subject to uniform national regulations. There is an emission standard of 1.2 pounds of pollutant per million BTUs for coal-fired plants and 0.8 pound for oil-fired plants. Partly in response to pressure from producers of high-sulfur fuels, the regulations were modified in 1978 to require stack scrubbers.[12] Both methods of reducing emissions—switching to more-expensive low-sulfur fuels or installing emission desulfurization technology—adversely affect measured productivity.

Frank M. Gollop and Mark J. Roberts have developed econometric estimates of the productivity impact of the EPA regulations. The framework of analysis they use is that of the translog cost function. The rate of productivity growth is equal, more or less, to the rate of decrease in costs, holding other things equal. The model also allows for biases in technical change, so that, for example, an increase in the cost of energy is assumed to alter the nature and pace of technology development.

The key variable is regulatory stringency, RS, which they define as a function of legally mandated reductions in emissions and the effective enforcement of these standards:

$$(4\text{-}1) \qquad RS_t = \left[\frac{E_t^* - S_t}{E_t^*} \right] \left[\sum_{i=t-1}^{t} \frac{1}{2} \frac{E_i^* - E_i}{E_i^* - S_i} \right].$$

In this expression, S_t, E_t, and E_t^* are all emission rates in pounds of sulfur dioxide per million BTUs and are, in turn, the legal standard applicable to the utility, the actual emission rate, and the utility's unconstrained emission rate. The importance of regulation is measured, therefore, as the product of a term reflecting the extent to which the utility is constrained by the emissions regulations, and a term reflecting the degree of enforcement.

12. For most producers the cheapest way to meet the regulation would be to switch to low-sulfur coal. But Congress mandated stack scrubbers, and once these are installed, the demand for high-sulfur coal is sustained.

This latter term is a two-year moving average of the ratio of the actual reduction in emissions to the required reduction.[13]

When entered into their cost function, the *RS* variable is highly significant. Regulation raised costs and lowered productivity for the coal- and oil-fired utilities in their sample. It consisted of fifty-six companies, twenty-four of which faced binding emissions regulation in 1973–74, a number that had risen to forty-one by 1978–79. According to the estimates, over the six-year period 1973–79, the companies facing constraints experienced an average annual reduction in their productivity growth rates of 0.59 percent. This number also matches fairly well with a difference of 0.48 percent in the rates of productivity decline for constrained versus unconstrained companies (−1.35 percent a year for the constrained and −0.87 percent for the others). The largest impact of regulation occurred in 1976, when growth was reduced by 2 percent among constrained firms.[14]

The Gollop and Roberts results look sensible. Regulation took about half a percentage point off the productivity growth rate of electric utilities after 1973. Other aspects of their modeling, however, give rise to concern. Their estimates suggest that there was negative technical change (technical regress) of more than 6 percent from 1973 to 1975. This finding is explained in the model by the bias of technical change. It was fuel-using in the power industry, they say, so that the sharp jump in fuel prices in 1973–75 caused the technical regress.[15]

This scenario is implausible. The long-run pattern in the power industry through 1963 was of improvements in energy efficiency. Between 1922 and 1963 energy efficiency rose 2.6 percent a year; technological developments in the industry were sometimes described as BTU chasing.[16] The period from 1973 to 1980 was actually one of deteriorating energy efficiency, despite the jump in fuel prices. Gollop and Roberts ignore the problems created by excess capacity. The sharp dip in utilization noted earlier is a much more plausible explanation of productivity collapse in 1973–75, and also helps explain the loss of energy efficiency.

Despite this reservation, the partial effect of regulation that they find is probably a good estimate. Their regulation variable makes sense and was

13. Gollop and Roberts, "Environmental Regulations and Productivity Growth," pp. 662–63.

14. Ibid., p. 670.

15. Ibid., pp. 669–70.

16. Loftness, *Energy Handbook*, p. 153. The data exclude nuclear and hydroelectric generation.

constructed with care. Its impact probably was not biased by other problems in the results. Their findings do apply only to fossil-fueled plants, however; they do not try to estimate the impact of regulation on nuclear power.

Innovation and Technology in the Power Industry

Problems created by excess capacity and government regulation account for only a small part of the productivity collapse in the power industry. In the textile industry a major technological breakthrough in the 1960s spawned a successful wave of innovative machinery in the 1970s. By contrast, in electricity generation there were technological breakthroughs in the 1950s and 1960s—namely supercritical turbine technology and the development of nuclear-powered generating plants—that failed in the 1970s.

FOSSIL-FUELED GENERATION. In independent studies, Robert Gordon and Paul Joskow have reached similar conclusions about the evolution of technology in coal-, oil-, and gas-fired plants.[17] Both drew on the engineering literature, on interviews with plant managers, and on cost and productivity estimates that separated out different facilities by type and age. Facilities are important because the basic technology of using fossil fuels to generate electricity uses the Rankine cycle, in which fuel is burned to generate high-pressure steam. The steam is then expanded through a turbine, which turns a generator to produce electricity. An electricity-generating facility has relatively few employees, so that labor productivity has not been a primary concern. Of greater concern is the thermal efficiency of the units.

For many decades the technological changes that led to higher thermal efficiency were associated with large units because, within limits, a larger boiler is more efficient. And for a given size of unit, efficiency is also increased by raising the temperature and pressure. Because of the thermodynamic properties of the Rankine cycle, however, there is an effective limit on thermal efficiency achievable by these means. As this limit was approached in the 1960s, it became harder and harder to make units more efficient.

17. Robert J. Gordon, "The Productivity Slowdown in the Steam-Electric Generating Industry," Northwestern University, February 1983; and Paul L. Joskow, "Productivity Growth and Technical Change in the Generation of Electricity," Massachusetts Institute of Technology, December 1985.

As the gains from scale were exhausted, an attempt was made to increase efficiency further by developing very high pressure, or supercritical, units. Based on theoretical calculations and prototype developments, this technology promised substantial productivity gains. The new technology involved the use of steam pressures of around 3,500 pounds per square inch, compared with earlier technologies using 2,000–2,500 psi. The first supercritical units appeared in 1960, but the new technology diffused slowly, becoming significant only in the late 1960s and early 1970s. By the middle 1970s, however, the industry began to move away from supercritical units because of serious reliability problems. Though these had shown up in early units, supporters of the technology had hoped the bugs would be overcome. Instead, users found that reliability problems were endemic to the technology and that attempts to overcome them drove up construction costs.

NUCLEAR PLANTS. Already viewed by many as the wave of the future, nuclear power became even more attractive when oil prices jumped in 1973.[18] Particularly in areas of the country where coal was unavailable, nuclear power offered substantial opportunities for fuel cost saving. Although only 53 nuclear plants were in operation in 1975, 190 more were already on order.[19]

Unfortunately, expectations of cheap electricity in a nuclear future proved wildly optimistic. The technology made stringent demands on its conventional and nuclear components, requiring them to perform under wide variations in operating conditions. Also, the complexity of the interaction of different systems virtually ensured that some potential problems would go unforeseen. This optimism translated into productivity terms in two ways: capital costs were much higher than expected and performance was much lower.

Capital costs are extremely important for nuclear power; in 1983 they were estimated to be two-thirds of the lifetime costs for a nuclear plant.[20] Overruns in construction costs have been tremendous. For example, one utility whose staff we interviewed completed a plant in 1985 costing $4 billion; its basically identical sister plant had been completed ten years earlier at a cost of $875 million. Using the price deflator for construction (where inflation has run significantly higher than the average), the difference is still a factor of two and a half.

---

18. This section was written by Robert Krebs.

19. U.S. Nuclear Regulatory Commission, *Annual Report, 1975*, pp. 173–85.

20. Charles Komanoff, *Power Plant Cost Escalation: Nuclear and Coal Capital Costs, Regulation, and Economics* (New York: Van Nostrand Reinhold, 1981), p. 14.

Cost overruns have frequently been blamed on longer construction times, but some analysts reject this view. Regression results reported by Charles Komanoff found that in the nuclear industry as a whole, cost escalation was not attributable to the increase in construction time.[21] Several caveats are in order here. For one thing, the period of his analysis was 1971 to 1978; since then construction times have become even longer. For another, Komanoff assumes that a longer construction time adds to interest costs but not to construction costs. Thus his estimates of construction costs ignore the opportunity cost of funds used; he does not inflate the yearly flows to reflect the potential earnings forgone.

Komanoff finds that changes in regulation played the most important role in capital-cost escalation. As different quirks in plant designs surfaced, the Nuclear Regulatory Commission required an increasing number of costly changes to be made in existing plants and those under construction. Particularly after the accident at Three Mile Island, the stringency of regulation of plant design and construction sharply increased.

Another problem in nuclear power generation is that the capacity factors projected for nuclear plants have never been realized. While the decision to build a plant was generally based on calculations assuming that it would produce 80 percent of its potential output and be available for baseload generation on a predictable basis, the average actual capacity for the industry from 1960 to 1982 fluctuated between 53 and 60 percent, and units were often shut down at unpredictable times.[22] As a rough estimate, this implies that the ratio of capital to output for nuclear plants is half again higher than originally predicted. Moreover, the larger plants, which have been built since the early 1970s, have performed more poorly than the industry average.

It is possible, in fact, that scale has had a significant role in the failure to generate electricity cost-effectively. According to Richard Hellman and Caroline J. C. Hellman, the optimism of the 1960s led utilities to build nuclear plants with four times the capacity of any previous nuclear plant and larger than any existing coal plant.[23] The problem for such plants was that their equipment operated at a lower pressure than was conventional in steam turbines (1,600 psi). To generate electricity on the scale of 1,200 or more megawatts required unprecedented volumes of steam and completely new steam generators and related systems. The Hellmans cite Victor

21. Ibid., chap. 10 and app. 3.
22. Richard Hellman and Caroline J. C. Hellman, *The Competitive Economics of Nuclear and Coal Power* (Lexington, Mass.: Lexington Books, 1983), p. xvii.
23. Ibid., p. 7.

Gilinsky, then commissioner of the NRC (from 1982), to the effect that the technology was introduced before the bugs had been worked out and that the technical issues were more complex than the utilities could be expected to manage. Problems were at first attributed to inexperience, and this was no doubt partially the case. However, Komanoff's regressions found that learning effects had little impact on construction costs, and that realized economies of scale were much more modest than had been predicted. Piecing this together with the generally poorer performance record of larger plants, it seems reasonable to infer that the large nuclear plant, a technological-engineering innovation of the late 1960s, was a failure in practice in the 1970s.

Conclusions

Of the four industries we studied, three (chemicals, machine tools, and electric power) had substantial slowdowns in productivity growth and one (textiles) had no slowdown. We have argued that the evolution of technology in these industries has been a vital part of their overall performance and productivity.

There was a sharp slowdown in both new product innovation and in productivity-enhancing process innovation in the chemical industry after 1973, matching its slowdown in productivity growth.

In the textile industry, by contrast, there was no slowdown in the main form of product-related innovation, namely, the introduction of new textile machinery. This again corresponds to the productivity performance of the textile industry.

The performance of the machine tool industry was also closely tied to its pattern of innovation. Starting from a strong technological base in the 1960s, the industry as a whole failed to keep pace with the speed of worldwide technological development. When it tried to play catchup after 1977, the rapid flow of innovations was not accompanied by rapid productivity growth because U.S. firms had been seriously weakened in several crucial respects.

Finally, the two key innovative technologies for electricity generation that were developed in the 1960s failed in the 1970s. This failure, along with the earlier development of excess capacity and environmental regulation, again tracks the productivity performance of the industry.

We also found distinct parallels between the experience of the chemical and electric power industries. Both suffered sharp declines in demand growth; both were slow to recognize this and then slow to adjust because of the long gestation periods in construction. Both used the benefits of larger and larger scales of production to achieve rapid productivity gains in the 1950s and early 1960s, and both industries found these gains had been largely exhausted by 1973 or before. In the chemical industry, there was a dearth of major innovations and stagnant productivity. Electricity generation was actually worse off, because major breakthroughs were tried and failed.

Before drawing together the lessons of the four industrial sectors, we turn to an analysis of the impact of technology outside the industrial sector. It is clear that for two decades the computer and electronics industries have been highly innovative and dynamic. What is not as obvious is the extent of the benefits achieved by this technological development in the using industries. The textile industry achieved productivity gains as a result of technology developed in the machinery industries. Why has the electronics revolution failed to induce a similar success?

CHAPTER FIVE

Electronics and White-Collar Productivity

DURING the past thirty years technological change in the electronics field has proceeded very rapidly.[1] That statement is self-evident, but if quantitative evidence were needed, it would be provided by recently developed price indexes. A new price index was constructed by the U.S. Department of Commerce for office, computing, and accounting machinery, based on the information-processing capacity of the machines. The index fell by about 9 percent a year from 1969 to 1973, and by 16 percent a year from 1973 to 1982, relative to the overall price index for nonresidential equipment.[2]

It is now possible to buy enormous information-processing power cheaply, and the business sector of the economy has been making the purchases. Today almost a third of every dollar spent on business equipment goes into information-processing and related equipment.[3]

In the preceding chapters, we have considered whether the opportunities for technical advance may have diminished, or if industries may have failed to take advantage of those that were there. The electronics revolution looks as if it provides important counterevidence against both of these possibilities. And this revolution looks big enough that it could even have offset any slowing of innovation elsewhere. Anyone who had been told back in 1965 of the dazzling advances ahead in electronics and of the spread of computers throughout the economy would certainly have forecast a period of substantial productivity growth ahead.

Clearly this has not happened. Instead, paradoxically, rapid electronics innovation and rapid adoption of the new equipment seem to have led to only a minor productivity payoff. Can the paradox be resolved?

1. Some of the ideas in this chapter were presented in Martin Neil Baily, "Productivity and the Electronics Revolution," *Bell Atlantic Quarterly,* vol. 3 (Summer 1986), pp. 39–48.
2. Data from the national income and product accounts.
3. Ibid.

Three Empirical Studies

One possible resolution is that the electronics revolution has been successful and has raised white-collar productivity tremendously, but the gains have been offset by productivity declines among blue-collar workers. Stephen S. Roach has argued that this does not seem to be the case. In his studies he classified workers as either information workers or production workers, based on data by occupation provided by the Bureau of Labor Statistics.[4] He then selected two sectors, the goods sector (manufacturing, mining, and construction) and the information-intensive sector (services, communications, wholesale and retail trade, finance and insurance). For each of the sectors he computed total output per white-collar, or information, worker and output per production worker. The results are shown in figure 5-1. Within both sectors, output per information worker has fallen or grown more slowly than output per production worker.

These figures plotted by Roach are striking and have rightly drawn press attention, but he has overinterpreted what they show. Clearly businesses are choosing to produce by methods that use more information workers, but this does not tell us that there are problems in the activity of information processing, as Roach suggests. It is possible that employing more information workers has allowed output per production worker to rise as companies substituted information workers for production workers. Roach's data show an interesting pattern that needs to be incorporated into an understanding of what has happened. But they do not yet identify the problem.

A study by Gary W. Loveman tried to demonstrate more directly the failure of information technology to generate the expected productivity benefits.[5] Loveman started by defining information equipment as computing machines, data bases, purchased software, document-generation (word-processing) equipment, and telecommunications, telex, and satellite equipment. He then looked at data on information technology equipment in fifty-five businesses from 1979 to 1982. The data were drawn from the Management Productivity and Information Technology survey,

4. Stephen S. Roach, "The New Technology Cycle," *Morgan Stanley Economic Perspectives,* September 11, 1985, pp. 1–11.

5. Gary W. Loveman, "The Productivity of Information Technology Capital: An Econometric Analysis," Massachusetts Institute of Technology, January 31, 1986.

Figure 5-1. *Output per Production Worker and per Information Worker, 1962–84*

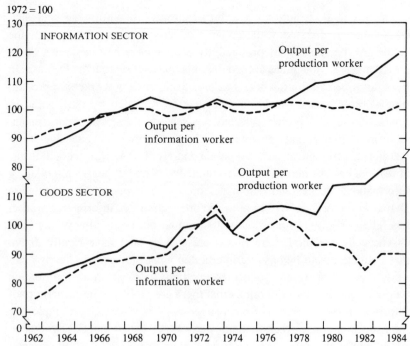

1972 = 100

Source: Stephen S. Roach, "The New Technology Cycle," *Morgan Stanley Economic Perspectives*, September 11, 1985, fig. 3.

which was made as an adjunct to the well-known Profit Impact of Marketing Strategy, or PIMS, data. The information allowed Loveman to estimate a production function for the fifty-five businesses, separating conventional capital (equipment, structures, and land) from information capital, such as computers and communications equipment. A business was defined as an operating division or a product group within a large company (mostly Fortune 500). All of them were in manufacturing.

The data set itself created many difficulties and problems. Capital measures were based on book value, price deflators were not specific to companies, and the calculation of labor input was somewhat crude. However, Loveman has both acknowledged these problems and made creative attempts to reduce their impact. His results reveal the existence of a smaller output elasticity for information capital in a Cobb-Douglas production function than the equivalent output elasticity of production capital. Loveman has interpreted this finding as evidence that information capital is less productive than other capital, although he is quite cautious about his results, noting that different specifications or estimation methods give somewhat different numbers.

Despite his caution, we believe that Loveman has also overinterpreted his results. It is not the output elasticity per se that reveals the productivity of a factor, but rather the relationship of the factor's marginal productivity to its cost. Information capital is likely to have a smaller coefficient in the production function of manufacturing enterprises than regular capital because less information capital is needed than regular capital. It remains to be seen, therefore, whether Loveman's data show direct evidence of the relative ineffectiveness of information capital.[6] Beyond this, his results are limited because his data set is restricted to manufacturing, while the white-collar productivity problem is concentrated in industries outside manufacturing.

The third empirical study was carried out by Paul Osterman, who analyzed the impact of computers on clerical and managerial employment over the period 1972–78.[7] Osterman specified a model in which the demand for clerical and managerial labor depended on the wage rates of these groups and on an estimate of the quantity of computers in use. His regressions were run on a cross-section of industries, and the employment figures were from the Current Population Survey. Computer use by industry was based on a Data Resources, Inc., survey and measured the total amount of main memory in each industry.

The basic results show that computers do reduce the demand for clerks and managers in the first instance. However, once Osterman takes account of the timing of the effects, most of the initial loss of employment following computer installation is eliminated by later hiring. He hypothesizes that in the long run, computers encourage companies to engage in more computer-related activities rather than reducing the labor required for a given level. This is basically the idea that we take up in the remainder of this chapter.

---

6. Gary Loveman has pointed out to us in a personal communication that in several of his regressions the coefficient on information capital is essentially zero. If this is the case, clearly the resulting estimate of the marginal productivity of this capital would also be zero.

7. Paul Osterman, "The Impact of Computers on the Employment of Clerks and Managers." *Industrial and Labor Relations Review*, vol. 39 (January 1986), pp. 175–86.

Measured, Social, and Private Returns

Data that are currently available do not reveal the productivity of information capital. Developing better data bases is an urgent priority for productivity analysis, and one we will pursue in later work. At this point, the best way to explore the puzzle we have raised in this chapter is through analysis and simulation. If it is true that electronics innovation has not yet paid off, we need to know what explanations are plausible and how they relate to standard economic principles. Three hypotheses stand out. One is that some output is not being measured. The second is that companies are engaging in activities that are privately, but not socially, productive. The third hypothesis is that there are incentives affecting the behavior of individuals that are not in the interests of organizations.

Measurement

This issue, which was first discussed in chapter 2, is important here because we think the electronics revolution may have increased unmeasured output. Today, even manufacturing companies provide important services to their customers. For example, computerized production techniques offer greater diversity in the design and configuration of items produced, allowing greater customer choice. Such services as information on how to use the product effectively may be provided for customers along with the sale of a physical product. Within the service sector itself, products have changed even more dramatically. Computerization in a bank allows customers to move money among accounts, select from many different types of account, and make a variety of portfolio transactions. Yet the system of price surveys run by the Bureau of Labor Statistics is not set up to capture either the service component of manufacturing production or the nature of changing output in the service sector. Since the price series do not capture the changing character of production, inflation is overstated and productivity understated. This is dramatically evident in the banking industry and a few others, where there is no attempt at all to capture the prices of services supplied. Real output is assumed to be in a fixed relation to labor input.

Distributional versus Productive Activities

It has long been recognized that advertising and marketing activities may be privately profitable but not socially valuable. Advertising has its biggest impact in persuading customers to choose one brand versus another, has a small impact in persuading customers to buy one product rather than another, and has little influence on overall spending decisions and overall output. Some argue that advertising makes markets work better by giving new products and new ideas a chance. But the bulk of marketing activities are not of this form. They are intended to persuade customers to shift from one established company to another.

In the reverse of the case of technology development, private returns may exceed social returns because marketing is inherently *distributional* rather than socially productive. It may be quite rational and efficient for market participants to invest resources to increase their own profits at the expense of the profits of others. Another sphere in which this issue arises is in appraising the value of assets. A company that buys another company that is undervalued, or even a division or a plant that is undervalued, may earn a higher rate of return than it would receive from buying machinery or developing new products.

The electronics revolution is a revolution in information processing. Undoubtedly, greater use of computing power allows plants to run more efficiently, and computers reduce the number of clerks needed to process accounts or payroll. But much of the information-processing technology is used as a tool to pry customers away from one's competitors. And some of it may be used to develop corporate strategies that at best yield capital gains at the expense of the original owners of the assets that have been rearranged. Innovation in electronics has made distributional activities much more attractive. The *New York Times* recently described "a broad industry trend in which computers are being viewed not merely as time-saving devices for such tasks as accounting, but as strategic weapons that can help a company outdistance its competitors."[8]

Individual and Organization

Economics emphasizes the role of markets in allocating resources. But in practice markets play only a limited role in allocation decisions within organizations. It is particularly difficult to establish internal markets for white-collar activities because information is so hard to value. The output may consist of a series of reports or a set of tables. The information con-

8. Andrew Pollack, "IBM to Offer Computerized Analysis of Systems," *New York Times,* September 29, 1987.

tained in these documents might be crucial to the company's success or might spawn a series of meaningless meetings.

The difficulty of valuing information interacts with another aspect of bureaucratic organization. Salary and position depend heavily on the number of persons being supervised. When technology makes information processing cheaper, there is an incentive for executives to generate more information, rather than reducing the staff required to complete the existing tasks. This problem typically does not arise in a manufacturing plant, where output and productivity are more easily monitored. A new generation of equipment is assessed on how much labor or material it saves.

This hypothesis and the measurement problem can be traced to a common element: the difficulty of valuing information and the services it provides for users. The Bureau of Labor Statistics finds it difficult to put a price on this part of output. Companies find it hard to allocate resources because they are unable to price the flows of information used internally.

All three hypotheses about the impact of the electronics revolution are linked to a kind of market failure. For markets to work well, the prices facing individuals or companies when they make decisions must be representative of the values of the products or services to the economy as a whole. When appropriate prices are not known or when decisions are based on private values that differ from social values, then productivity, particularly measured productivity, will suffer.

A Model of the Effect of Electronics Innovation

To explore two of these ideas more fully, we have developed a model that embodies both a source of measurement error and a gap between the private and social returns. The issue of the gap between individual incentives and organizational goals is not discussed further here.

A company produces a well-defined product that is sold in the marketplace. It could be a physical commodity or a service like a trade on the stock exchange. The quantity produced of this product is Q, the price it is sold at is P, and both of these magnitudes are known. Output is produced using labor (L), production capital (K), and information capital (IK).

$$(5\text{-}1) \qquad Q = F(L^Q,\ K,\ IK^Q).$$

The superscripts on labor and information capital indicate that they are used in the production of the measured or regular output Q.

The company also provides a service to its customers. This service, S, is produced by labor and information capital, with the obvious notation.

$$(5\text{-}2) \qquad\qquad S = G(L^S, IK^S).$$

The service is provided free to customers, and is assumed to be of genuine value to them.

Third, the company produces another form of output, one that we will call marketing (M), with a production structure similar to that for customer services.

$$(5\text{-}3) \qquad\qquad M = H(L^M, IK^M).$$

The key element of the marketing activity is that it simply moves industry demand around without adding to the total. The outputs of the three activities of the company are all included in the demand curve facing this company.

$$(5\text{-}4) \qquad\qquad Q = D(P, S, M, P^{ind}, S^{ind}, M^{ind}).$$

The demand for the regular product of this company depends on the price it charges, the amount of service provided free with the product, and the amount of marketing it does. Demand decreases if the firm raises its price, both because existing customers buy fewer units and because some customers switch to other companies. Demand increases with either an increase in the amount of service provided or an increase in marketing. The demand for the product of this company also depends on the average levels of price, service, and marketing set by other companies in the industry. When the average price in the rest of the industry increases, it raises demand for this company. When the level of service provided by other companies in the industry increases, or when the average marketing effort made by other companies increases, it lowers demand for this company. In all of these cases, these are straightforward competitive responses, assuming that customers are looking around for the best deal and that marketing activities do influence customers' choices.

There is an important difference between service and marketing. Marketing is a kind of zero-sum activity. An increase in marketing by all firms together leaves overall demand unchanged.

This model is not intended to capture strategic behavior. The company makes its own decisions (for P, S, and M and levels of factor use) assuming that the industry variables are fixed. The company ignores any response of the industry to its own actions. In the end, however, the company finds that other firms in the industry set price, service, and marketing at the same level as it has. Other companies in the industry are like this company.

The main conditions that are necessary if the company is maximizing its profit are:

$$(5\text{-}5) \qquad P(1 - 1/E) = P^L/F_L = P^K/F_K = P^{IK}/F_{IK}.$$

The conditions are that the marginal revenue product (E is the elasticity of demand for the company's own product) is equal to the marginal cost of production for each factor, where P^L is the price of labor, F_L is the marginal product of labor, and the same notation holds for capital and information capital.

Customer service and marketing affect profit indirectly by allowing the company to raise its selling price for a given quantity of sales. The conditions that determine how many workers and how much capital are used in service and marketing activities are:

$$(5\text{-}6a) \qquad (PQ/S)(E_S/E) = P^L/G_L = P^{IK}/G_{IK},$$

$$(5\text{-}6b) \qquad (PQ/M)(E_M/E) = P^L/H_L = P^{IK}/H_{IK}.$$

These conditions relate the elasticities of the demand curve with respect to service and marketing to the marginal productivities of the labor and information capital used to produce them.

To carry the analysis further we substituted in Cobb-Douglas production functions for regular output, service, and marketing; and a constant-elasticity demand curve. This resulted in four sets of parameters. The first set consisted of three alphas in the regular production function, applied to labor, capital, and information capital. Constant returns to scale is assumed in this function. There are two betas in the production function for service and two gammas in the one for marketing. In the demand function the final set of parameters consisted of three deltas for the company's own decision variables (price, service, and marketing) and a further three deltas for the industry values for the same variables.

With these substitutions, the model can be solved, and the first element in the solution determines the way in which the company chooses labor, capital, and information capital to produce regular output. The solution to the model, in this case, sets the shares of sales revenue (PQ) devoted to each of the factors of production, given as follows:

$$(5\text{-}7a) \qquad P^L L_Q/PQ = (\alpha_1/\delta_1)(\delta_1 - 1),$$

(5-7b) $$P^K K/PQ = (\alpha_2/\delta_1)\,(\delta_1 - 1),$$

(5-7c) $$P^{IK} IK_Q/PQ = (\alpha_3/\delta_1)\,(\delta_1 - 1).$$

Under perfect competition, the shares of total revenue being paid to the three factors would be the alphas from the production function. But for this company, the price elasticity of demand (δ_1) modifies this result and means that total revenue is not exhausted by expenditures on factors used directly in production.

The shares of revenue allotted to the labor and capital used to produce customer service and marketing depend on their production elasticities (the betas and gammas) and on the elasticity of demand for these intangibles relative to the basic price elasticity (δ_2 and δ_3 relative to δ_1).

(5-8a) $$P^L L_S/PQ = \beta_1(\delta_2/\delta_1),$$

(5-8b) $$P^{IK} IK_S/PQ = \beta_2(\delta_2/\delta_1),$$

(5-8c) $$P^L L_M/PQ = \gamma_1(\delta_3/\delta_1),$$

(5-8d) $$P^{IK} IK_M/PQ = \gamma_2(\delta_3/\delta_1).$$

There is an important asymmetry between the decisions made about regular output and the decision to devote resources to marketing activities. A company with monopoly power will set price above marginal cost, restrict output, and hurt consumers. But on the other hand, a monopolist will not engage in marketing activities and will not incur the deadweight loss resulting from the attempt to move customers around. If the level of marketing activity everywhere doubles (M and M^{ind} double), this does not shift the demand function, but will result in a higher price and a lower output.

The level of customer service that would be provided in a competitive situation is not set in this model. If the company were perfectly competitive (δ_1 and δ_2 become infinite), then the level of service (S) is indeterminate. The existence of monopoly power indicates that the company will keep service output below its optimal level, but this may or may not mean that the *service intensity* of output is higher or lower than under competition, since output is also restricted.

The Effect of a Decline in the Price of Information Capital

The model does not allow for innovation that occurs directly within the industry we are studying. The assumption is that the semiconductor, com-

Figure 5-2. *Baseline Simulations of Output, Service, Marketing, Information Capital, and Price of Output*

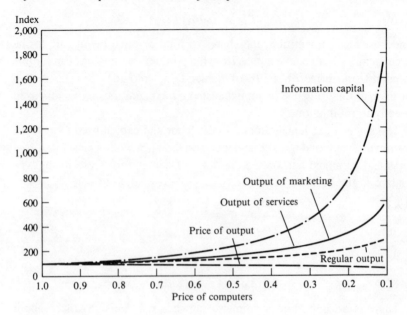

Figure 5-3. *Baseline Simulations of Labor and Capital Productivity*

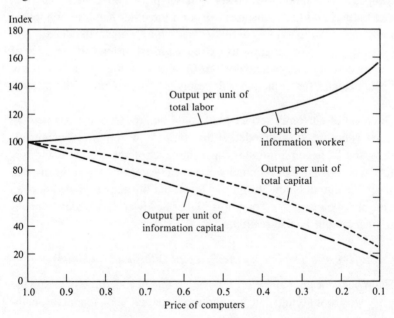

puter, and related industries develop the new technology, while the industries that use the equipment benefit from rapidly declining prices. We use the model now to simulate the impact of a 20 percent a year decline in the price of information capital for ten years. We experimented with different parameter values for the alphas and deltas and so forth, and use the following as central reference values. Labor has a 60 percent share in the production of regular output. Labor and information capital have 50–50 shares both in the production of customer service and in marketing. The own price elasticity of demand is 3, while the demand elasticities of service and marketing are both 0.2. There is little to guide our choice on these last values, but we find that if these values are raised by much (to 0.4) the model no longer reaches a stable solution. If these values fall, the effect is fairly predictable. The importance of the service activities gradually diminishes. The industry price elasticity is unity, the industry customer service elasticity is 0.1, and the industry marketing elasticity is 0.2. The main results, shown in figures 5-2 and 5-3, reveal:

—The decline in the price of information capital results in much more intensive use of this factor, both in production and in the provision of services and marketing.

—Because they use information capital more intensively than regular production, the output of customer services and marketing rise more than regular output.

—Labor productivity rises and capital productivity falls as the substitution of information capital for labor takes place.

—There is no difference in the changes in output per production worker and output per information worker, even though customer service and marketing are not included as part of output.

The results of the simulation model, therefore, do follow some aspects of the electronics revolution. The use of information capital grows rapidly, capital productivity tends to fall, and information-related activities such as customer service and marketing expand rapidly. A prediction of the model that is surprising and does not accord with observation is that there is no change in output per production worker relative to output per information worker. The existence of unmeasured output, even combined with the rapid decline in the price of the principal factor producing that output, is insufficient to induce a decline in output per information worker relative to output per production worker.

Two features of the model may explain this result. The first possibility is that we assumed information capital contributed to the production of the regular measured output as well as producing customer service and marketing. Thus cheaper information capital resulted in the substitution of capital for labor and raised labor productivity across the board. To explore this possibility, we ran a new set of simulations in which information capital did not contribute at all to the production of regular output. Production capital, K, gets the full share of 0.4 in the basic production function, and information capital gets a share of zero. The results of this simulation are shown in figures 5-4 and 5-5.

In this variant of the basic model it is notable that there is now almost no decline in the price of the basic product resulting from the decline in the price of information capital. Customer service and marketing activities increase tremendously and then increase output and sales even without a price change. The labor productivities remain completely flat in this variant of the model for both full and partial measures of productivity. Varying the model eliminates overall productivity growth, but even though the uncounted output increases, the substitution of information capital for information labor prevents a decline in output per information worker.

We believe that this variant of the basic model helps in understanding the characteristics of the model, but clearly it does not completely explain the puzzle of declining output per information worker. We suggest, therefore, a second variant of the basic model. We reinstate the original basic production function, allowing information capital to contribute to the production of regular output, but we assume that there are constant elasticity of substitution (CES) production functions for both customer service and marketing activities, and that the elasticity of substitution between information labor and information capital is less than unity. We have explored various values of the elasticity and report the results for a 0.2 elasticity in figures 5-6 and 5-7. *Once the elasticity of substitution is set low enough this version of the model tracks the observed trend of declining output per information worker.*

Lessons from the Model

The electronics revolution has put cheap computing power in the hands of companies buying the new equipment. This technology has not been enough to generate rapid gains in productivity for the economy as a whole and has even been combined with declines in output per information

Figure 5-4. *Baseline Case, but No Contribution of Information Capital to Production*

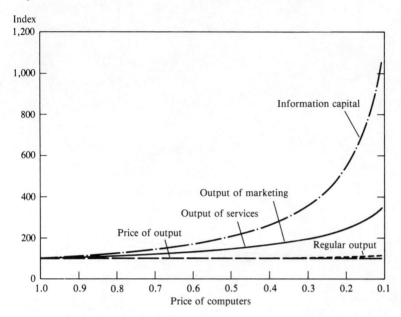

Figure 5-5. *Labor and Capital Productivity with No Contribution of Information Capital to Production*

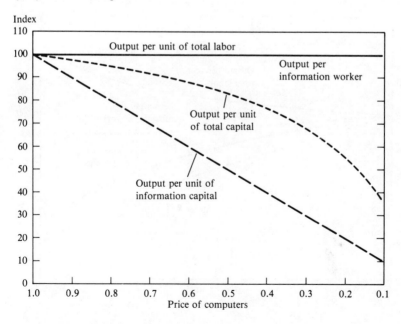

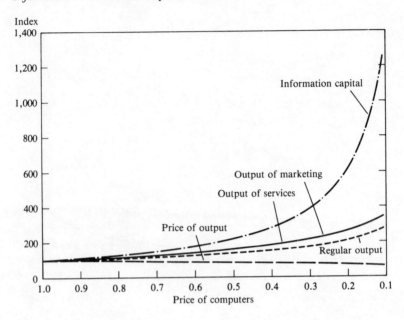

Figure 5-6. *Baseline Case, but Low Substitutability between Information Labor and Capital*

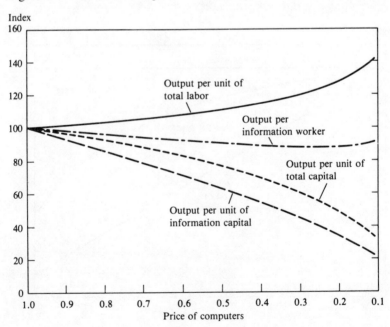

Figure 5-7. *Labor and Capital Productivity with Low Substitutability*

worker. We have explored a simulation model in which companies provide not only a regular product, but also customer service and marketing activities. Customer service provides real value to consumers, but this value is not counted in today's standard output measurements. Marketing activities are purely distributional. Starting out, we thought these characteristics in themselves would be enough to construct a model that was consistent with the puzzling productivity weakness associated with the electronics revolution. But in a model in which information capital contributes to the production of regular output and in which information capital can be readily substituted for labor in all activities (Cobb-Douglas), this turned out not to be the case. Companies step up service and marketing activities, but cheap information capital allows capital-labor substitutions that boost productivity across the board.

One possible way to reconcile the model with observation is to argue that information capital contributes little or nothing to regular production. And indeed when that assumption is incorporated, productivity stagnates. The fact that the computer revolution has affected the office more than the factory, or staff functions more than line functions, may be part of the story. However, even in the extreme case of a zero contribution of information capital to regular production, the model registers no decline in output per information worker, so that more is required for it to be consistent with observation.

One assumption that does provide the desired consistency is that substitution possibilities between information capital and information labor are limited in the service and marketing activities. In this case, the company offers more services and steps up marketing because of cheap information capital, but then has to hire more workers to operate all the computers it has purchased.

We believe these results point to an overall hypothesis about the productivity paradox associated with the electronics revolution. The primary impact of the revolution has not taken place in the factory or in the production of directly observed and measured output. Because of this, the revolution has failed to boost overall productivity by much. The electronics revolution has encouraged companies to engage in activities that provide real but unmeasured output, or else are socially unproductive. Computers and other equipment are not readily substitutable for information labor, so production requires more and more information workers as the price of computers falls.

Conclusions on White-Collar Productivity

We have discussed three hypotheses to explain why white-collar productivity has stagnated. To these we could add one more. The transition to a new technology is always costly and sometimes productivity declines before it rises. The transition to a computerized economy has involved a major shift in technology, and it may be taking several years to adjust to the change. Office personnel have to be retrained, and a new system often seems to go down endlessly before it eventually generates high productivity.

If this last hypothesis is correct, it may not be necessary to do anything more to solve the problem. Future productivity gains should be much greater than those of the recent past. We hope this is the case, but are rather dismayed by the persistence of the slowdown in the business sector as a whole.

If measurement is a problem, then the Bureau of Labor Statistics and the Commerce Department should launch an initiative to bring their price surveys more in line with the reality of a service-oriented economy. Important policy decisions have to be made about productivity, and these decisions should be based on a sound understanding of the facts.

If markets are providing too much incentive for moving customers or assets around, the solution is hard to see. Restricting private sector activities would not be appropriate. The best solution, we argue later, is to raise the incentives for productive activities and use them to shift the market balance toward growth.

CHAPTER SIX

Lessons from the Studies
and Strategies for Growth

IN A SUPERB article, Henry Ergas has drawn lessons about the effect of technology on economic growth based on the experience of the major economies of the Organization for Economic Cooperation and Development.[1] He distinguishes between two strategies. One pushes out the technological frontier and develops leading-edge industries; the other concentrates on diffusing technology that has already been developed. The latter approach seeks to incorporate into existing industries the developments that have been made in leading-edge industries. Germany and other European countries, he claims, have been highly successful in the second strategy of refining and perfecting existing technologies, while Americans have been extremely successful at pushing out the frontier but only partially successful at diffusing and applying technology. The British have made many innovative developments in technology but have failed badly at the task of diffusion and refinement. The Japanese have primarily followed a strategy of borrowing and refining technology, but in recent years they have gradually changed their strategy to include the development of certain leading-edge industries as well.

Ergas argues that increases in productivity depend most heavily on success in diffusing and refining technology, and as a result the Japanese and some Europeans have enjoyed high productivity growth.[2] The reasons he

1. Henry Ergas, "Does Technology Policy Matter?" in Bruce R. Guile and Harvey Brooks, eds., *Technology and Global Industry* (Washington, D.C.: National Academy Press, 1987), pp. 191–245. The cross-country analysis of Richard Nelson points in similar directions. See Richard R. Nelson, *High-Technology Policies: A Five-Nation Comparison* (Washington, D.C.: American Enterprise Institute, 1984).

2. Ergas's discussion of France is problematic for his ideas. He claims that France followed a strategy of leading-edge development, not diffusion. This, he says, explains why Germany has had more success than France. But in productivity terms, France has done as well as or better than Germany. If Ergas's overall framework is correct, as we believe it is, then France must have achieved considerable success in diffusion from 1955 to 1979. The French must be closer to the

gives for these conclusions are, first, that new ideas and technologies as they emerge are applied only to small segments of the economy. Thus even if a country has a monopoly on revolutionary technological developments, the early payoff will not be all that large for productivity, jobs, and trade. And second, after a few years it is relatively easy for other countries to copy the early breakthroughs. Indeed, some argue that coming second confers certain advantages, because it is easier for the copiers to see where the new technology can be developed and applied.[3] In answering the question posed in the title "Does Technology Policy Matter?" Ergas concludes that policy is much more important if it can help make good use of available new ideas than if it encourages the creation of those ideas.

Although Ergas does not extend his analysis to a consideration of the post-1973 slowdown in productivity growth, this extension can readily be made for the slowdowns in Europe and Japan. After the end of World War II, the United States held a substantial technological lead. Not only did it lead in the development of new technology, it also started out as the most productive economy by a wide margin. Though Britain and France challenged the U.S. technology lead in certain industries, Japan and most of the countries of Europe followed a strategy of diffusion and catch-up. They became very good at it and grew quite rapidly until their leading industries rivaled U.S. industries in levels of productivity. At that point, they had to slow down because the frontier of technology was growing slowly.

In this framework, the slowdown in productivity growth in Japan and Europe was inevitable because the rapid productivity growth of earlier years had been achieved by closing the gap with the United States.

Technology and the U.S. Slowdown

The slowdown in the United States does not follow from this line of analysis with quite the same inevitability, but the U.S. slowdown becomes much more comprehensible if one follows the same line of thinking. The most important element in Ergas's cross-country comparison is that taking advantage of available technology is more important to growth than cre-

---

Japanese case than Ergas states. We are not experts on the French economy, but it may be that the French situation looks worse today than it did a few years ago, and this recent experience colored Ergas's analysis.

3. For a discussion of this issue, see David J. Teece, "Capturing Value from Technological Innovation: Integration, Strategic Partnering, and Licensing Decisions," in Guile and Brooks, *Technology and Global Industry,* pp. 65–95.

ating new technology. And our research strongly supports the idea that productivity slowdowns took place in the case-study industries because these industries failed to take advantage of technology that was potentially available. Slow growth in the United States has reflected missed opportunities.

The United States was the pioneer in computer technology and continues to hold a substantial lead there. This edge created a built-in opportunity for the U.S. machine tool industry to become the world leader in computer-controlled machinery, which, in turn, could have helped many other manufacturing industries boost their own productivity. These opportunities were missed.

As the pioneer in computers, the United States also had the opportunity to reduce labor in white-collar activities by seeking ways to take advantage of the new technology. Although economists lack, as yet, a full understanding of the behavior of white-collar productivity, it is clear enough that streamlining labor needs in service activities has failed to provide the source of growth that it might have.

The United States was the pioneer in nuclear energy and made important developments in supercritical turbine technology. These developments in applied science provided opportunities for innovation, but the new technologies were not implemented successfully and did not result in productivity improvement.

The U.S. chemical industry found that its sources of growth had been exhausted, but there were advances in materials science, in biotechnology, and in the manipulation of molecular structures that could have opened up new opportunities.

The one U.S. industry we studied that sustained rapid productivity growth was the textile industry, and this was the one clear example of an industry that did take advantage of the new technology that became available to it.

To summarize: for productivity to keep growing, there have to be new ideas and the economy has to take advantage of them. Japan and Europe increased productivity not only because advanced technology was available to be borrowed, but also because they became adept at the activity of borrowing and even at improving on what they borrowed. As they approached the frontier, Japan and Europe faced an inevitable slowdown.

The slowdown in the United States was not inevitable in the same way. It occurred more because the U.S. economy failed to incorporate new technology effectively into production than because the scientific frontier stopped expanding. That conclusion is the key lesson from the case studies in this book, and it fits with the key lesson Ergas draws from his cross-country comparison of technology policies.

The behavior of the manufacturing sector in the United States since 1981 adds support to this analysis. The rise in the value of the dollar after 1980 was a serious blow to the tradable goods–manufacturing sector. The sector responded to this by rapidly boosting productivity. Their quick response suggests that the potential for growth had existed for some time but companies took advantage of it only under the pressure of trade. The textile industry is notable in this context because it faced such pressure earlier and responded earlier. If technological opportunities had in fact been permanently depleted, the manufacturing industry could not have responded to the high dollar shock as it did.

Why Were Technological Opportunities Missed?

Opportunities were missed, we suggest, for four main reasons. The first is that incentives in the United States were adversely affected when other countries borrowed U.S. ideas.

The Adverse Effects of Being Caught

It is all very well for Ergas to recommend that countries follow a strategy of diffusion and catch-up as a way of increasing productivity growth, but not everyone can play catch-up. Such a strategy implicitly relies on someone, notably the United States, playing a disproportionate role in pushing out the frontier. The entire game is dangerous because the very success of the countries that are catching up erodes the return to the pioneer. Even when the United States had a clear lead in technology, inventors and innovators received only a fraction of the social return from their activities. Once other countries made up the gap in technology, U.S. innovators found many additional foreign competitors poised to take advantage of whatever they did.

This experience has created a particular disadvantage for U.S. industries. The Japanese and Europeans did not have to put their best people and their greatest managerial effort into developing brand new ideas. They could borrow what they needed and put their efforts into refining products and into process improvements, including better ways of managing labor and managing inventories. The United States has not developed its ability to diffuse and refine technology, relative to other industrial countries. Opportunities for growth were missed because the United States developed a comparative advantage in creating opportunities, but a comparative *dis*advantage in exploiting them.

Management Failures

Foreign competition cannot be blamed for all of the missed opportunities. In fact, in the past few years it is the nontradable sector that has suffered the biggest productivity problem. The collapse of productivity in electric power was not the result of foreign competition. Similarly, the failure to boost white-collar productivity is a domestic problem.

In the review of explanations for the slowdown in chapter 2, we discussed the hypothesis of managerial failure. We pointed out a central difficulty of blaming managers for the slowdown: why should they suddenly have gotten worse? Is it the case that managers around the world suddenly suffered a collapse of decisionmaking power?

The case studies and the general hypothesis about the slowdown we are now describing does revive the issue of managerial failures. And we want to reiterate the idea that we do not believe that managers have been infected by a stupidity virus manifesting itself in bad decisions. Instead, we argue that the challenges they faced became different and harder. The technology frontier was still moving out after 1973, but the opportunities this movement presented were much harder to exploit than they had been in earlier decades. For example, in the 1950s and 1960s raising productivity in electric power was fairly easy; in general, managers simply exploited the well-known benefits of increased scale. By contrast, managing the demands of supercritical turbine technology and nuclear power in the 1970s and 1980s presented much more difficult tasks. Similarly, managing the computer revolution presented tremendous opportunities for failure as well as opportunities for success. Most managers did not understand the technology very well and were often at the mercy of technical staffs who were not tuned to the larger goals of their companies.

Weak Demand

Successful innovation requires both that a technological opportunity be available and that companies see market conditions as favorable to its introduction. In a major study of innovation, Jacob Schmookler has argued that strong growth in demand is the key factor in rapid innovation.[4] He contends that innovation is induced by the potential for profitable development. If the profit potential exists, argues Schmookler, there will be ways of making technological breakthroughs. To support this hypothesis, Schmookler points to evidence that in areas where rapid changes in technology took place, a surge in patents *followed* a surge in industry output.

The Schmookler view could explain why innovation has been weak in the 1970s even though opportunities have been available. Supply shocks in the 1970s, particularly those linked to food and energy prices, boosted inflation in the United States and elsewhere. Policy, particularly monetary policy, responded to the higher inflation by constraining the growth of aggregate demand. From 1970 to 1985 the average civilian unemployment rate in the United States was 6.9 percent, much higher than the average over earlier periods.[5] Weak aggregate demand, therefore, might have discouraged companies from taking advantage of innovation opportunities.

Our case studies lend some support to the Schmookler idea. For example, although the basic polymer for nylon was developed in 1934, during the depths of the Great Depression, the waves of innovations and patents that exploited the breakthrough did not roll in until the more favorable conditions of the postwar period. In the chemical industry in the 1970s, it is highly plausible that excess capacity dampened process innovations. Process-technology improvements generally increase the yield of a plant without increasing the capital and labor used in production. Since the industry was operating with chronic excess capacity, the incentive to devote resources to such innovation was weak.

The case of the electric power industry is somewhat similar to that of the chemical industry in this regard. Persistent excess capacity in the industry meant that investment for expanding capacity was low or nonexistent. Under these conditions, the manufacturers of turbines had little incentive to invest in solving the problems of supercritical turbine technology. And some of the same is true for efforts to solve the problems of nuclear technology.

Schmookler's argument and evidence, plus the support given to them by our case studies, lead us to accept the hypothesis that weak demand did

4. Jacob Schmookler, *Invention and Economic Growth* (Harvard University Press, 1966).
5. Data are for civilian workers. *Economic Report of the President, January 1987*, table B-31.

adversely affect innovation in the 1970s. But we are not prepared to make weak demand a central cause of missed opportunities.

We are reluctant for two reasons. First, the Schmookler hypothesis has come under fire because of the potentially deceptive nature of his timing evidence. A high rate of patenting reflects the spread or development of a breakthrough in technology, not the breakthrough itself. The surges in patenting that Schmookler observed actually followed important technological developments. For example, the wave of electronics patents and innovations followed several years after the development of the transistor and the printed circuit. A major innovation takes place that causes product prices in an industry to decline. Demand increases, stimulating further technological improvements that then generate a rapid flow of patents. A criticism of Schmookler, therefore, is that technology is the driving force despite the observed tendency of patents to lag behind demand growth.

Second, the evidence from our own case studies fails to point to an unambiguous response of innovation and productivity to demand. For instance, we found that the pressure of foreign competition as it cut into the domestic demand for textiles actually stimulated the U.S. industry to *adopt* new technology as a survival tactic. This example seems to generalize to the manufacturing sector as a whole, as we noted earlier, because the pressure created by the strong dollar after 1980 apparently stimulated productivity growth.

Private Returns and the Effect on Productivity

One of the themes of this book has been that the return to the innovating company may not match the returns to society and hence private returns may not match the impact of innovation on productivity growth. This divergence, we have argued, can go either way, depending on the circumstances. The circumstances in the past fifteen years may have been such that investments in new technology that did not contribute to measured productivity rose relative to investments in new technology that did. One reason for this has already been discussed. Japan and the countries of Europe and Southeast Asia, using borrowed technology, have caught up to U.S. industries. Consequently, the returns from some traditional, productivity-enhancing forms of innovation in the United States may have been eroded because U.S. companies appropriate fewer of the returns from these innovations. The second reason formed the subject of chapter 5 on the electronics revolution. In traditional manufacturing, the productivity effect of innovation normally occurs, to a great extent, in downstream or using industries. In electronics, these downstream effects either have not taken place or have not been measured, because innovation in computer technology has taken the form of cheaper information processing, encouraging marketing activities or unmeasured increases in customer service.

Conclusions on Technology

It may well be the case that in many industries the technological frontier moved out more slowly from 1973 to the present than was the case in the 1950s and 1960s. But the results of this study, which seem to fit with conclusions drawn from cross-country comparisons, indicate that the biggest problem in the technology area lies in the failure to take advantage of opportunities and surmount the challenges these opportunities present.

Strategies for Spurring Growth

Improving the rate of U.S. productivity growth substantially may require concerted efforts in several directions. Certainly the analysis in this book has pointed to problems with worker skills, capital investments, managerial failures, and a variety of other areas that have contributed to the slowdown. But it is not clear that policy can or should necessarily try to solve all these problems. In any case, we ourselves want to concentrate on three policy issues in the technology area. First, should the United States government change the level of its support of basic research or restructure the way funds are allocated? Second, are there areas of technology development (middle-ground research) that merit additional support? And third, is there a case for maintaining or increasing the tax support provided for commercial R&D?[6]

In framing answers to these questions we want to acknowledge a lesson learned both from Ergas and from Richard Nelson. Technology policies may work very differently in different countries because of unique cultures and institutions. For example, the diffusion strategy in Japan has worked exceptionally well, operating through an elite government bureaucracy

6. The discussion of these issues draws on work carried out by Baily for the National Science Foundation and for Coretech, an industry-university consortium. In the latter case Baily worked with Robert Z. Lawrence of Brookings. Neither organization has sponsored this study and neither has influenced the recommendations given.

whose members work with various industries to inform them about and then encourage them to adopt the best available new technologies. A somewhat similar strategy was also attempted in Britain. Government agencies worked vigorously to bring about the rationalization of several fragmented industries and to encourage the adoption of technological change. Yet the results have been largely disastrous.

The most important institutional fact about the United States is that among the major industrial economies, it is the one that relies most heavily on the market to achieve growth. For most of its history this reliance has been a vital ingredient in U.S. economic success. But it is also the case that market incentives for technology development have been found to be inadequate in certain ways.

Public versus Private Support of Technology

There is a legitimate concern among applied researchers that the United States is missing opportunities in the technology area that other countries are taking advantage of, either because they are better able to cooperate on new technologies, or because there is more public support for technology, or both. At the same time, there is legitimate concern among policymakers that more public support may waste taxpayers' money and more public involvement may distort decisions and result in commercial failures. Everyone is afraid of letting the government pick winners and losers.

No single answer can resolve these concerns. A successful technology policy will involve several institutional arrangements, depending on the technology concerned. To determine what the different approaches should be, it is worthwhile clarifying some of the strengths and weaknesses of public and private projects.

The market generates a diversity of approaches to new technology and can find solutions that might not have been considered or even thought of by a centralized decisionmaking body. For example, when energy prices rose in 1973 and 1979–80, it was impossible to know where new conservation technologies might show up and whether new sources of energy would be made economically viable. The government-planning approach included a synthetic fuels project, which turned into an expensive failure. The market's combination of new conservation technologies and new products that use less energy proved remarkably successful in raising overall energy productivity.

It is harder to kill a project that fails when its support is provided by the government. A company will continue to fund an ongoing project only if the expected payoff from the *increment* to funding justifies the cost. Government projects often suffer from the tyranny of sunk costs because decisionmakers do not want to admit failure and may not be around when the final accounting on the project is made. This difficulty of killing projects is intensified when the inputs to a project become valued for themselves rather than for their contribution to developing a commercially viable product or process.

In a study of government-sponsored commercial technology development currently being undertaken for Brookings, Linda R. Cohen and Roger G. Noll review a series of disasters, including the synfuels project and the breeder reactor, that were killed too late. The Clinch River breeder reactor in Tennessee is notable because it was perpetuated by its ability to create jobs and profits in Tennessee.

Government-sponsored projects seem to work best when the goals are technical or mission-directed. Examples include the development of the atomic bomb and the effort to place a man on the moon. Private-sector projects work best in focusing their goals effectively on commercial viability and the usefulness of the results to the economy at large.

The private sector is reluctant to take on projects that are large and risky even though the expected payoff may be quite large. In principle, a market economy should be able to solve the problem of risk by spreading it across many agents. For a single company, a large research project might entail too much risk. But if the owners of the company are, say, pension funds holding hundreds of different stocks, then the risks to these owners of even a large project are minor. In practice, however, there is a separation of ownership and control in companies. The only way shareholders can judge the performance of managers is by the performance of the company, often the very short-run performance. The impact of external factors or the results of an unlucky event cannot be separated out. The managers of a corporation suffer severe economic consequences if the corporation fails or has a serious setback. The managers make the decisions and they are averse to risk.

Suppose a project costing $5 billion over five years has a 50 percent probability of yielding large returns and a 50 percent probability that the money will be lost. Suppose the expected rate of return is 60 percent, much better than the average return on investment in the U.S. economy. The size of the stake for society is only 0.02 percent of GNP a year (a trivial amount, the equivalent of someone with $50,000 a year risking $10), so the project looks worthwhile. However, no private company alone would commit $5 billion with a 50 percent chance of failure, so the market fails to provide an adequate incentive for risky projects.

In principle the government should be able to bear risk better than the private sector can, acting as the representative for the whole society, for whom the collective risk is small. In practice such projects are difficult for the government also. It is hard to decide whether a particular government-sponsored project is a worthwhile example of the government's absorbing risks, or whether it is a politically motivated effort to pump money into a state or a congressional district.

The most important difficulty the market has in providing incentives for technology development is the appropriability problem. This problem, which occurs when only part of the *social* return to R&D becomes a private return to the company that has paid the bill, has been mentioned in earlier chapters of this book and is widely acknowledged. But the magnitude of the difficulty it creates is often not sufficiently appreciated. Consider first the case of commercial R&D, the activity where the appropriability problem is the *least*. The social return to commercial R&D is two or even four times the private return. This means that, if a private company invests in R&D, it is analogous to setting up a new factory and then discovering that its competitors have the right to use the plant free for a half or three quarters of the year. Or it is like selling a product but receiving only a half or a quarter of the purchase price. The gap between the private and social rates of return to R&D reflects an externality that imposes a huge implicit tax on technology development. This tax creates the presumption that not enough R&D is being performed.

The patent system does help companies to appropriate more of the returns from their work. But patents certainly do not solve the problem. The gaps that have been measured between private and social rates of return occurred in the presence of patents.[7]

The appropriability problem not only results in too little commercial R&D, it also means there is little or no private sector funding for basic research or for research in the "middle ground." Companies choose to perform R&D only in those areas where they see an adequate private rate of return. When Edwin Mansfield and others estimate the gap between

---

7. Patent policy deserves a separate analysis. Tightening patent protection helps alleviate the appropriability problem, but at the expense of worsening the distortions created by the nonrivalry aspect of technology. Ideally, the discoverer of new technology should be rewarded, but the new technology should be made widely available.

private and social rates of return, their estimates are based on only the limited selection of projects for which companies had anticipated adequate private returns. Many projects with high social returns are simply never considered by the private sector.

The need for public support for basic research has long been recognized and substantial funding is supplied in this area. The same is not true for research in the middle ground. The absence of any clear institutions to support middle-ground research in the United States is a serious deficiency of current technology policy. Middle-ground projects consist of applied research that has commercial applications, but where the results are too general to make them attractive to private companies on their own. For example, Bell Labs was an institution performing important middle-ground research before the AT&T breakup. Essentially, it was publicly funded (through the telephone rate base) and privately operated.

These descriptions of the ways in which private and public support for technology succeed and fail pose a challenge for technology policy to maintain the advantages of the market while supplementing the incentives it provides. It is impossible to meet this challenge fully, but it is possible to do quite a bit better than we do now.

Support for Basic Research

In 1987 the federal government provided about $7 billion (in 1982 dollars) for basic research, a figure that has grown 3.8 percent a year since 1977.[8] Almost half of this support was for the life sciences, particularly those related to health. Most public funds for basic research go to universities.

There is general agreement that the United States has the strongest basic research community of any country in the world and that this preeminence has not diminished over time. If anything, such yardsticks of achievement as the number of Nobel Prizes awarded support the view that the U.S. domination in basic science has grown.

Public support of basic research in the United States should be considered a success, and the success stems from two sources. First, projects are selected on the basis of the internal logic and opportunities of the scientific disciplines themselves. Basic research is not motivated by a market result, so that the absence of market information as an input to the selection

8. National Science Foundation, *Science and Technology Data Book, 1987*, NSF 86-311 (Washington, D.C., 1986), pp. 7, 11.

process is not a disabling problem. Second, even though the money comes from the government and is disbursed through government agencies, there is still a high degree of decentralization in actually choosing topics. Peer review is used intensively to rate proposals and to decide on areas that need support.

Since the system works well it is hard to justify tinkering with it, but we are going to suggest that anyway. The United States has had trouble taking full advantage of its strength in basic science, and some of the responsibility for that may come from the scientific community itself and not just the people doing the applications. One reason that the U.S. scientific community fails to contribute more to economic success is that pure scientists are too attached to the linear model of technology development, a model that enshrines basic science as an isolated source of new knowledge that needs no stimulus from practical problems.

Stephen J. Kline has described the dangers of using a linear model for analyzing technology development.[9] A linear model implies that pure scientists work on knowledge for its own sake, the knowledge is then developed in practical ways, and finally the R&D people take the ideas or inventions and develop commercial products. But this model is incorrect, says Kline. Problems and questions that arise both in applied research and in R&D require solutions from basic research, so that commercial R&D can stimulate basic research. In addition, basic research can benefit from results generated in applied research. There is a need for interaction among people in all three activities.

In some disciplines, the flow of people and knowledge among practical applications and pure research is already large. Medicine is a prime example, and the progress made in this discipline reflects the advantages of frequent interaction. In other disciplines there is less contact with practical issues, so that peer review results in an academic community that is too self-referenced. Research is done to rebut someone else's research, not to push forward the discipline.

Some support for basic research can be justified on the basis of a search for pure knowledge. But the amount that taxpayers are willing to provide on that basis is small. In practice, Congress supports basic research because of its contribution to economic growth. Thus it is appropriate that applied scientists and R&D directors provide information about their needs

---

9. Stephen J. Kline, "Research, Invention, Innovation, and Production: Models and Reality," Report INN-1 (Stanford University, Department of Mechanical Engineering, February 1985).

and interests to the National Science Foundation and other agencies when the funding strategies for basic research are being set.

Support for Middle-Ground Research

The need for more effort in middle-ground research has been apparent for many years, but little has been done to achieve this goal.[10] However, the U.S. economy's faltering ability to compete in world markets has created a new urgency in Washington, and the time may be ripe for change. There are two avenues to increased middle-ground research. One is private: alter market institutions so that firms themselves generate more research. The second is public: provide more government funding.

During the Reagan administration government policies aimed at altering market institutions have already undergone some changes, and more are contemplated. Currently the favored approach is to reduce the appropriability problem by allowing coalitions of private companies to do joint research projects. The private rate of return to technology can be raised, so the argument goes, if the costs of a project can be split two ways or several ways. The Justice Department has significantly eased its antitrust policies to allow joint-venture work, and it has reduced its opposition to oligopolies with large market shares. The National Cooperative Research Act of 1984 put congressional approval on joint ventures. At present, the main difficulty with joint ventures from the antitrust point of view is not the Justice Department, but rather a fear that competitors excluded from a joint venture may file damage suits.

Of course, free-market lawyers and economists in the Justice Department have always been uncomfortable with using intervention as a way of achieving competition in the market. In their view, only clear, explicit price-fixing behavior justifies a legal response. A company or oligopoly that achieves a dominant market share through successful management should not be penalized. In a climate of deregulation, this view has flourished. Another reason for the department's shift is that the U.S. economy has become much more exposed to foreign competition. In the 1960s one could argue that GM's dominant market share and cost advantage presented at least the danger of a monopoly in the absence of antitrust restrictions. Such concern would hardly be justified today. Even if GM were the only domestic manufacturer, it would still be operating in a very competitive automobile market.

10. The issue is discussed in Richard R. Nelson, Merton J. Peck, and Edward D. Kalachek, *Technology, Economic Growth, and Public Policy* (Brookings, 1967).

In our view it was appropriate to relax the restrictions on coalition or joint-venture research and it may be appropriate to make other changes of the same type, for example, improving patent protection for intellectual property. But we are concerned that changes like these really fail to solve the problem of generating support for middle-ground research.[11] A specific example will illustrate the problem. A group of semiconductor manufacturers has proposed forming a joint-venture enterprise called Sematech. They plan to make advances in process technology that fall under the heading of middle-ground research. Once the new processes have been developed, each company will individually apply them to reduce production costs. Our concern is that since several companies will have the technology, the pressure of competition will tend to drive the private returns from the project close to zero for all the participants. *Joint-venture research projects do result in lower costs for each participating company, but they may well lower the private returns also, possibly to zero.* As a result, many companies are reluctant to join research coalitions. They say they are not interested in supporting the creation of new knowledge that their competitors will learn about too.

Joint ventures actually create game theory problems for companies. A given company can join, knowing its competitors will learn the same things it learns. It can refuse to join and risk a situation in which it is excluded from the knowledge that others will learn. And to complicate the picture, a company can decide how much of its existing technological knowledge to put on the table in a joint venture. Companies face a "prisoner's dilemma" when they join a research coalition.[12] If all of the companies reveal all their relevant existing knowledge, the project has the best chance of succeeding. But individual companies have an incentive to withhold information and hope that others will reveal their knowledge.

These difficulties explain why there is not widespread pressure today to form research coalitions. IBM, for example, would generally be better off going it alone. The Sematech proposal grew out of concern that the Japanese industry had already moved ahead in technology, so that unless the U.S. industry caught up, the companies would forfeit the return to their existing physical and knowledge capital. And of course the Sematech proposal includes a request for cofunding by the government.

11. We have benefited from discussions with Charles Herz and Robert Anderson of the National Science Foundation on this issue.

12. The prisoner's dilemma occurs when two people commit a crime together. The police have no clear evidence, but place the criminals in separate cells. If both refuse to confess, they get off free. If one confesses first, he draws a short sentence. The one that holds out gets a long sentence if his partner has confessed first.

Some observers argue that the Japanese economy has been able to solve the problems inherent in stimulating middle-ground research by balancing cooperation and competition. Coalitions are formed to perform middle-ground research and then the partners divide up the results. One company applies the technology to one set of new products and another company is directed to develop a different set of products. In this way the participants all get at least a temporary monopoly on certain aspects of the new technology, and this temporary monopoly provides the necessary private returns. In practice, the Japanese track record is mixed. Japanese firms have been highly successful in gaining market share in semiconductors, but they seem to have driven everybody's private returns down to the point where the return on physical capital is barely adequate. So far, they have been unsuccessful in making inroads in the computer industry.

Clearly the obstacles to stimulating middle-ground research are steep. Given U.S. market institutions, the best available strategy is to encourage private coalitions with a well-defined role for public participation. Because of the difficulty companies will have in earning an adequate rate of return, it is appropriate to provide part of the cost with public funds. Because of the dangers of pork-barrel projects, it is appropriate to insist that most of the funding be private. Moreover, the projects should never become government projects. The impetus must come from the private sector, and the decisions about what to work on and how to proceed should be made by the private sector. Company scientists should perform the research. There is a role for a publicly designated ombudsman or trustee, who would monitor the project to make sure, if possible, that the participants did not succumb to the temptations of the prisoner's dilemma and to make sure that the project's public funds have a reasonable chance of providing general benefits to the economy.

This strategy has no guarantee of success, but it is a way of maintaining the market's advantages while closing somewhat the gap between private and social returns.

Tax Credits for R&D

The presumption in favor of tax support for R&D is extremely strong. The externality in the production of knowledge implies that the private sector will underinvest in this activity. A variety of econometric evidence indicates that private companies do respond to changes in economic incentives. These arguments mean that there is a clear case for tax support for R&D unless the weight of the specific evidence goes against it.

Critics of the existing tax credit make four arguments.[13] First, they argue that public funding would be better directed to basic research. Second, they say the existing tax credit has flaws in its design. Third is the concern that companies have taken advantage of the credit by exaggerating their eligible expenditures. And fourth, there is some evidence that the credit has failed to stimulate R&D. We consider these arguments in turn.

BASIC RESEARCH VERSUS R&D. We have not advocated any cutting back on funding for basic research. On the contrary. But the general conclusions of this book are that the slowdown has not resulted from any weakness in basic science. Indeed, it appears that it is precisely in the area of the commercialization of technology that the United States has had problems relative to other countries. There is no case at the present time for shifting resources out of R&D into basic research. We disagree with the critics on this issue.

THE DESIGN OF THE EXISTING CREDIT. When the credit went into effect in 1981 it was a 25 percent credit on eligible R&D expenditures above a base level. The base level consisted of expenditures over the preceding three years. Policymakers hoped that this credit would reduce the marginal cost of performing R&D by 25 percent and hence provide a serious incentive for companies to do more of it, even though the expected revenue loss from the credit was small.

In fact, however, the credit failed to provide a 25 percent incentive because increases in R&D in one year entered the base-level calculation in the three later years and reduced the credit in those years. The rolling base meant that the effective incentive of the credit was much smaller than 25 percent. Early estimates of its actual effect were incorrect, because people neglected the interaction of the credit with the corporate income tax. The effective incentive of the credit from 1981 to 1985 was 7 to 8 percent. But because the rate of the credit was later dropped to 20 percent, and because the marginal rate of corporate income taxation was also dropped, the effective incentive in force today is only about 4 percent. This covers the 1986–88 credit.

There are two responses to this criticism. The first is to point out that

13. These criticisms were raised, for example, in the testimony of Robert Eisner, professor of economics, Northwestern University, in *Research and Experimentation Tax Credit,* Hearings before the Subcommittee on Oversight of the House Committee on Ways and Means, 98 Cong. 2 sess., August 2, 3, 1984, pp. 71–141.

even though the existing credit provides only a small incentive, it also costs very little. When the credit provided a 7 to 8 percent incentive, it cost $1 billion to $1.5 billion a year.[14] A flat rate subsidy of this size on $50 billion of corporate R&D would have yielded only a 3 percent incentive. Today's incentive is lower, but the cost is also lower, well under $1 billion.

The second response to the issue of the incentive effect of the credit is based on a study for the National Science Foundation by Charles River Associates. They have evaluated the incentive and proposed a different design for the credit.[15] If a company's base were increased each year by some exogenous factor, such as the overall rate of inflation, this adjustment would restore much of the original intent of the credit and provide a marginal incentive that was closer to the statutory rate.[16]

In brief, we agree with those critics of the tax credit who say that its current impact is small, but we disagree with the conclusions they draw. Unhappily, the credit has failed to provide the marginal incentive that it was expected to, yet it has provided an incentive that is quite good relative to its revenue cost. Nevertheless, the credit could be larger and it could be redesigned so that R&D expenditures in one year do not reduce the credit received in later years.

RECLASSIFICATION OF EXPENDITURES. Some economists argue that companies responded to the introduction of the credit by looking for ways of boosting eligible R&D expenditures without boosting their actual R&D effort. Evidence for this view can be found in the claims companies made for the credit on form 6765 for 1981.[17] These data show that the average rate of increase in eligible R&D expenditures was much higher than the rate of growth of R&D reported in other sources. In addition, ineligible expenditures actually fell in 1981 relative to 1980.

It seems clear that these data do show that companies reclassified expenditures. But the implications of this for the validity of the credit are limited. First, the problem of reclassification really only applied importantly in the first year the credit was introduced. After that, the divergence

14. Ibid., pp. 71–74.

15. Charles River Associates, *An Assessment of Options for Restructuring the R&D Tax Credit to Reduce Dilution of Its Marginal Incentive,* report prepared for the National Science Foundation, Division of Policy Research and Analysis (Boston, February 1985).

16. Such a redesign would also deal with another issue. With the current design of the credit, a company with flat or falling R&D expenditures actually has an incentive to reduce its R&D in order to lower its base in later years. A redesign would eliminate this problem.

17. See Eisner testimony, in *Research and Experimentation Tax Credit.*

in rates of expenditure growth fell off. Companies did at first look for ways to boost their eligible spending, but they could not keep adding activities to boost their spending growth rates each year. And, ironically, the boost in expenditure in 1981 from reclassification actually cost the Treasury almost nothing. Because of the way the rolling base was introduced, the present value of the tax saving achieved by a company that reclassified its expenditures in 1981 was almost zero.

Second, the data supplied by the U.S. Office of Tax Analysis from the 6765 forms are based on unaudited returns. Because the eligibility rules were still unclear in 1981, companies claimed everything they thought might go through. But the tax return of every major corporation is audited, so the final level of approved credit claims was undoubtedly much smaller.

Third, eligibility rules for a tax credit can be worked out only by experience. Since 1986 the rules for the credit have been tightened to the point where only expenditures devoted to hard technology development are included. The reclassification problem is no longer an important issue.

THE EFFECTIVENESS OF THE EXISTING CREDIT. The impact of the credit that was introduced in 1981 has been analzyed in many empirical studies. This evidence has been surveyed by Jane G. Gravelle for the Congressional Research Service, Kenneth Brown for the Joint Economic Committee, and Joseph Cordes for the National Academy of Science.[18] All three surveys conclude that the credit did boost corporate R&D spending, but all three also agree that the effect was small. Since we argued earlier that the credit from 1981 to 1985 provided only a small incentive for R&D, we certainly do not question these overall conclusions. However, we do think that the credit's impact may have been somewhat larger than the surveys have suggested.

The evidence on the credit consists of two types. One type is derived from overall estimates of the responsiveness of R&D spending to the cost of performing it. And this evidence generally finds a substantial effect for the credit. The elasticity of R&D spending to cost is estimated to be between −0.3 and −1.0. The other type of evidence consists of specific studies of the impact of the credit, and a study by Eisner and a survey by

18. Jane G. Gravelle, "The Tax Credit for Research and Development: An Analysis" (Washington, D.C.: Congressional Research Service, January 25, 1985); Kenneth M. Brown, *The R&D Tax Credit: An Evaluation of Evidence on Its Effectiveness,* prepared for the Joint Economic Committee, 99 Cong. 1 sess. (Washington, D.C., 1985); and Joseph J. Cordes, "The Impact of Tax Policy on the Creation of New Technical Knowledge: An Assessment of the Evidence," George Washington University, February 1987.

Edwin Mansfield, in particular, conclude that the credit has had a small effect.[19] Mansfield surveyed about 800 companies and asked them retrospectively how much they would have reduced their R&D spending in the years 1981, 1982, and 1983 in the absence of the credit. The responses show a small but rising effect that indicates that the credit raised 1983 spending by 1.2 percent, relative to what it would otherwise have been.

We take Mansfield's survey seriously, but like many other economists, we are skeptical of the ability of surveys to discover the impact of a small credit.[20] For any individual company, the specific attributes of its potential projects are going to bulk larger in its decisionmaking than the tax credit. That does not mean the impact of the credit is unimportant in total. Moreover, we found in our own company interviews that R&D directors tend to attribute increases in their R&D budgets to their own successful research efforts. Corporate planners give more weight to the credit because it is one of the factors weighed by a company's board in approving a given budget.

The Eisner study is interpreted by him to show that the credit is ineffective, but actually the evidence he finds is quite mixed. First, the data do show that companies that applied for the credit have a much higher rate of increase in their R&D spending than the overall rate of increase for all companies. He interprets this as evidence for reclassification, as we noted earlier, but the same evidence is also consistent with there having been a real increase in R&D spending by eligible companies.

The same source of data, the Office of Tax Analysis, also provides a comparison between companies that had enough income to make use of the credit and those with insufficient income. This comparison shows no evidence of a difference in R&D behavior between the two groups. The comparison cannot be given too much weight, however, because companies in the sample were able to carry the credit forward or back, so most of them could potentially use it. And since the data cover only companies that applied for the credit, presumably most of them planned to use it.

Kenneth Brown has pointed out that Eisner in his testimony to the House Ways and Means Committee made a similar comparison of com-

19. Edwin Mansfield, "Statement to the House Ways and Means Committee on the Effects of the R and D Tax Credit," University of Pennsylvania, undated (Mansfield did not actually testify); and Eisner testimony, in *Research and Experimentation Tax Credit*.

20. Gravelle, "Tax Credit," p. 21, points out the problems inherent in a survey like Mansfield's.

panies with or without eligibility for the credit by using data from the comprehensive Compustat file. His data do indicate substantial differences in R&D spending between companies with full or zero eligibility for the credit. In 1983, R&D spending by companies fully eligible for the credit rose 30.4 percent, whereas spending by those ineligible rose only 11.1 percent. In 1982 the comparable figures were 29.5 percent and 19.6 percent, and in 1981, when the credit was only partially in effect, they were 39.5 percent and 33.8 percent.[21] We ran the same test for a sample of 172 firms from the Compustat file for 1985 and found that those eligible for the credit raised their spending 10.3 percent, compared with a decline in spending of 5.5 percent by firms that were ineligible.[22]

Finally, both Robert Eisner and we ourselves have run some cross-sectional regressions using Compustat data and found results that are largely inconclusive. We decided that it is impossible to isolate a clear-cut impact of the credit on R&D spending. However, it is also hard to find a plausible or stable relationship between R&D and other variables. The data are "noisy" so that it is hard to identify not only the impact of the credit against the static inherent in the data, but also the impact of other explanatory variables.

Our conclusion from the available evidence is that the influence of the R&D tax credit remains uncertain. There is quite a bit of evidence showing that R&D responds to incentives. And some of the evidence that has been cited as showing little effect actually shows potentially a substantial effect. Because of this mixed and inconclusive evidence, we decided to examine for ourselves the time-series data on R&D spending. The impact of the credit may have become more evident than it was in earlier studies as additional years of data have become available.

R&D Spending over Time

The simplest direct evidence on the effectiveness of the credit is to look at whether R&D spending grew more rapidly when the credit was in place than it did in earlier periods, taking into account prevailing economic conditions.[23] Tables 6-1 and 6-2 show what happened. From 1980 to 1985

---

21. Eisner testimony, in *Research and Experimentation Tax Credit*, table 9.
22. Our calculations were based on R&D adjusted for inflation.
23. Robert Z. Lawrence participated in this analysis and Greg Hume was the research assistant.

Table 6-1. *Growth of Company-Funded R&D Spending, by Industry and Industry Shares, 1960–85*[a]

Industry	Share of total R&D performed by all industries in 1980 (percent)	Growth rate over periods shown (percent per year)		
		1960–75	*1975–80*	*1980–85*
Lumber and furniture	0.5	[b]	3.3	−1.9
Stone, clay, and glass	1.3	2.5	2.5	−0.5
Primary metals	2.1	1.4	−0.5	0.6
Fabricated metals	1.8	2.4	3.3	−1.9
Nonelectrical machinery	18.6	6.3	7.1	7.8
Electrical machinery	19.3	5.0	6.9	9.3
Transportation equipment	24.7	5.2	8.7	3.0
Instruments	8.7	8.7	14.0	10.5
Miscellaneous manufacturing	1.2	7.0	3.7	8.3
Paper	1.8	n.a.	7.5	7.8
Chemicals	15.1	3.0	3.7	10.9
Petroleum	5.0	2.1	9.4	4.3
All industries	100.0	3.6	7.0	6.4

Source: Authors' calculations based on data from the National Science Foundation.
n.a. Not available.
a. R&D spending is in constant dollars.
b. Lumber and furniture did almost no R&D in 1960. Its growth rate over 1960–75 was 43 percent per year.

R&D spending grew at 6.4 percent a year, with growth strongest in chemicals, instruments, and electrical machinery.[24] The growth rate is almost twice the rate achieved between 1960 and 1975, but is slightly lower than the rate between 1975 and 1980, the period immediately before the credit was enacted. However, the evidence in table 6-2 provides a much better picture of what happened to the R&D effort of U.S. industry after the credit was introduced. Companies have to allocate funds to R&D out of their production revenues. The ratio of R&D spending by an industry to its production volume measures its commitment to R&D. And the rate of growth or decline in that ratio is a good measure of whether an industry is becoming more or less R&D-intensive.

The last two columns in the right-hand portion of table 6-2 show that the ratio of R&D spending to output during the period when the 1981–85 R&D tax credit was in effect grew more than twice as rapidly as it did in the comparable period before enactment of the credit. Judged by the ratio

24. The credit went into effect in 1981. The figures shown in table 6-1 for 1980–85 are the averages of the growth rate in 1980–81, 1981–82, 1982–83, 1983–84, and 1984–85—the relevant ones for the credit.

Table 6-2. *Ratio of Company-Funded R&D Spending to Industrial Production Revenue, Selected Years and Periods, 1960–85*[a]

Industry	Ratio expressed as an index (1980 = 100)				Average annual growth rate of the ratio (percent)		
	1960	1975	1980	1985	1960–75	1975–80	1980–85
Lumber and furniture	n.a.	108	100	72	[b]	-1.5	-5.6
Stone, clay, and glass	n.a.	109	100	88	0.2[b]	-1.7	-2.3
Primary metals	104	106	100	116	0.9	-1.1	3.2
Fabricated metals	106	105	100	86	-0.6	-1.0	-2.
Nonelectrical machinery	93	108	100	120	1.1	-1.4	3.6
Electrical machinery	166	126	100	110	-1.6	-4.1	2.7
Transportation equipment	73	83	100	92	0.9	4.1	-1.6
Instruments	80	88	100	130	0.7	2.7	6.8
Miscellaneous manufacturing	n.a.	98	100	130	2.4[b]	0.5	6.1
Paper	n.a.	95	100	120	n.a.	1.1	4.0
Chemicals	187	108	100	130	-2.8	-1.5	5.9
Petroleum	89	76	100	130	-1.0	6.4	5.9
All industries	99	94	100	120	-0.3	1.3	3.0

Source: Authors' calculations based on data from the National Science Foundation and the Federal Reserve Board.

n.a. Not available.

a. R&D spending is in constant dollars.

b. The ratio for lumber and furniture grew at 40 percent a year in 1967–75. The data for stone, clay, and glass began in 1962; for miscellaneous manufacturing they began in 1962. See note b to table 6-1.

of R&D to output, American industry is 20 percent more R&D-intensive than it was in 1980, and this surge is pervasive, with gains exceeding 10 percent in eight out of twelve industries. The table also shows how unusual the 1980–85 period was relative to the entire time from 1960 on.

Regression Analysis of Aggregate and Industry Data

We used regression analysis based on a simple model in which R&D spending in a given year depends on R&D in previous years, on output in the current and the previous years, and on two variables to capture the effect of the R&D tax credit.[25] This specification then allowed us to test whether the credit increased spending, and if so, by how much. This was tested separately for each of the manufacturing industries shown in tables 6-1 and 6-2 and for all the industries together. The data consisted of annual observations of company-funded R&D collected by the National Science Foundation for each industry.[26] The estimation was made first using a linear relation, for which the estimates showed the effect of the credit on the level of R&D spending in 1982 dollars. Table 6-3 displays these results. They show that from 1982 to 1985 the tax credit increased R&D spending in eleven out of the twelve industries. Only lumber and furniture had a negative coefficient. The sum of the coefficients indicates that the credit added almost $2.9 billion dollars (in 1982 dollars) to total R&D on average over the years 1982 through 1985. This figure is highly significant statistically. The credit was in effect for only half of 1981 and so we expect it to show up less strongly, and it does. Still, the results show a positive impact in all but three of the industries and a sizable total effect of $1.8 billion.

The final row of table 6-3 shows the results of estimating the same specification on data that aggregate all the industries. In principle, this calculation should give the same results as the line above for the sum of all industries. And, in practice, it comes pretty close. The average effect from 1982 to 1985 drops slightly, to $2.6 billion. The first-year effect drops more, to $1.3 billion.

25. Two dummy variables were used, one equal to unity in 1981, the first year when the credit was partially in effect. The other was equal to unity in 1982–85.

26. For some early years for some industries there is no breakdown of R&D into its federally and company-funded components. We extrapolated for these years using the available figures. Since the industries in question did little federal R&D, the errors made must have been very small. Industries where the data were seriously incomplete were discarded.

Table 6-3. *Effect of the R&D Tax Credit on R&D Spending, by Industry, 1981 and 1982–85: Linear Specification*[a]
Millions of 1982 dollars

Industry	Effect of credit in 1981		Average annual effect of credit, 1982–85	
Lumber and furniture	−6.9	(0.31)	−9.7	(0.63)
Stone, clay, and glass	32.4	(1.5)	16.9	(1.0)
Primary metals	71.6	(2.6)	43.9	(1.8)
Fabricated metals	25.9	(0.55)	10.0	(0.35)
Nonelectrical machinery	−77.0	(0.14)	704.2	(1.6)
Electrical machinery	313.9	(1.2)	626.0	(2.6)
Transportation equipment	637.1	(1.8)	586.1	(1.9)
Instruments	46.5	(0.48)	19.8	(0.14)
Miscellaneous manufacturing industries	59.5	(3.9)	93.2	(5.2)
Paper products	−10.2	(0.24)	14.8	(0.36)
Chemical products	535.5	(4.0)	701.8	(3.5)
Petroleum refining and extraction	173.4	(1.4)	71.9	(0.36)
Sum of all industries[b]	1,802	(2.4)	2,879	(4.3)
Estimate of aggregate equation	1,306	(1.3)	2,649	(1.9)

Source: Calculated from National Science Foundation data. Numbers in parentheses are t-statistics.

a. The estimated equation was $R\&D = \text{Constant} + a_1R\&D(-1) + a_2R\&D(-2) + a_3IP + a_4IP(-1) + a_5Credit81 + a_6Credit82\text{-}5$. The $Credit81$ variable was defined as unity in 1981 and zero otherwise. The $Credit82\text{-}5$ variable was defined as unity in 1982-85 and zero otherwise. The coefficients a_5 and a_6 are reported above (with their t-statistics).

b. The estimated t-statistics for the sum of the coefficients were computed assuming that the disturbances in the individual industries were independent of one another.

Table 6-4 gives the equivalent results when a logarithmic specification is used. This specification assumes that the credit changes R&D spending by a given percentage rather than by a dollar amount. Only two industries have perverse signs for 1982–85 and neither is significant. Overall, the credit had a strong positive effect. It increased R&D spending a little more than 7 percent, a figure that corresponds well to the results of table 6-3. The increase in R&D generated in 1981 was about two-thirds the size of later effects. Again, this matches the previous results.

The aggregate equation shown in table 6-4 gives a slightly stronger result than that obtained by averaging the individual industry results.

Taken as a whole our statistical investigation of time-series data on R&D expenditures by industry supports the view that the credit has increased R&D spending. The effect is extremely pervasive, affecting almost all the industries covered. It is quite substantial in magnitude, given the small size of the credit and the acknowledged problems that the credit has because of its moving company-specific base. Yet the magnitude is not

Table 6-4. *Effect of the R&D Tax Credit on R&D Spending, by Industry, 1981 and 1982–85: Logarithmic Specification*[a]
Percentage change

Industry	Effect of credit in 1981		Average annual effect of credit, 1982–85	
Lumber and furniture	−4.4	(0.16)	−3.6	(0.19)
Stone, clay, and glass	7.4	(1.4)	4.1	(1.0)
Primary metals	10.2	(2.4)	7.5	(1.9)
Fabricated metals	4.2	(0.47)	1.4	(0.25)
Nonelectrical machinery	−13.8	(0.25)	−0.34	(0.003)
Electrical machinery	3.8	(0.86)	7.6	(2.1)
Transportation equipment	7.2	(1.4)	5.8	(1.5)
Instruments	8.1	(1.7)	14.7	(2.6)
Miscellaneous manufacturing industries	13.9	(2.7)	20.7	(4.0)
Paper products	−1.2	(0.20)	2.8	(0.50)
Chemical products	10.7	(3.0)	12.3	(2.5)
Petroleum refining and extraction	13.4	(1.4)	12.3	(0.87)
Weighted average of all industries[b]	4.8	(0.80)	7.4	(1.8)
Estimate of aggregate equation	4.7	(1.4)	8.4	(2.0)

Source: Same as table 6-3. Numbers in parentheses are t-statistics.

a. The estimated equation was $\ln R\&D = \text{Constant} + a_1\ln R\&D(-1) + \ln a_2 R\&D(-2) + a_3\ln IP + a_4\ln IP(-1) + a_5 Credit81 + a_6 Credit 82\text{-}5$. The *Credit81* variable was defined as unity in 1981 and zero otherwise. The *Credit82-5* variable was defined as unity in 1982–85 and zero otherwise. The coefficients a_5 and a_6 are reported above, multiplied by 100.

b. The estimated t-statistics for the weighted average of the coefficients were computed assuming that the disturbances in the individual industries were independent of one another.

so large as to be implausible. The credit represented a subsidy of about 7 percent. Thus our results indicate that the elasticity of response of R&D was about unity. This is at the high end of the range of existing elasticity estimates, but not out of the range.

One possible explanation of the increase in R&D spending that took place between 1981 and 1985 is that it was caused not just by the credit, but by other factors. In particular, some companies may have stepped up their own R&D spending in the hope of winning defense contracts. Others may have responded to the pressure of increased foreign competition. However, the analysis of R&D by industry described above did not indicate that industries that were heavily involved in defense research showed above-average increases in R&D over the credit period. Nor were the increases in R&D restricted to industries under competitive pressure. Consider the transportation equipment industry. This industry includes automobiles (facing intense competition) and aircraft and missiles (where the defense department is the biggest customer). The industry increased its

R&D by 5.8 percent from 1982 to 1985, less than the average estimate for all industries. Some of the research and development in the automobile industry is considered ineligible for the R&D tax credit by the Internal Revenue Service, so the behavior of this industry is consistent with our conclusion that it was the credit and not other factors that made the difference. In general, the pervasiveness of the impact across industries supports the interpretation that the credit was working.

We also ran an additional regression to check further on the possibility that our tax-credit variables might be picking up an increase in R&D that would have occurred anyway because of competitive pressure. We added to our aggregate equation an additional dummy variable for 1978–80. About that time, U.S. companies started to realize that R&D was essential if they were to remain competitive. We did find evidence of a small surge in spending (about 3 percent) starting in 1978. But our estimate of the size of the impact of the R&D credit changed very little. The estimated stimulus in 1981 was 3.1 percent, rising to 6.4 percent in 1982–85.

Conclusions on the Credit

It is hard to make a case against tax support for R&D. Even if we were to accept the results of Mansfield's survey and conclude that the 1981–85 credit raised R&D only 1.2 percent, this increase still translated into an additional $500–600 million a year of corporate spending. And since the social return is two to four times the private return, this made the credit a good investment for taxpayers. We believe that the evidence supports a view that the credit affected R&D spending more than Mansfield's survey suggested. The credit added as much as $2.9 billion to corporate R&D spending, according to the time-series results. For any estimates between Mansfield's and this one, the social return to the credit was excellent.

The 1986–88 credit is smaller and its impact will also be smaller. In addition, its impact in the future will not necessarily show up in faster growth of R&D spending, because the long-run effect of a credit is to raise R&D-intensity, not to raise the long-run growth rate of research and development.

Weighing the presumptive arguments for tax support for R&D, the sum of the available evidence, and the potential for improving the incentive effect of the credit, we conclude that a continued or even an expanded credit for R&D is an appropriate part of technology policy.

Conclusions on Strategies for Growth

The current economic crisis in the United States is associated with a large gap between the rate at which spending has been expanding and the rate at which production has been expanding. According to our spending behavior, we are living in the 1960s with growth of 4 percent a year. But production is not being generated at a 1960s rate. Production is growing only 2.8 percent a year. The nation refuses to accept the consequences of the productivity slowdown for living standards. The gap between production and spending is being met by foreign borrowing, but this cannot continue much longer because the interest costs are skyrocketing. Eventually the gap must be closed, and though it will have to be closed in part by reduced spending, as far as possible it should be closed by higher productivity.

In a time of budgetary crisis it may seem quixotic to suggest maintaining funding for basic research, increasing funding for middle-ground research, and maintaining or increasing the tax support for research and development. But we believe that such pro-growth policies are worthwhile. In the long run they will help to solve the budget crisis and to close the gap between spending and production.

Index